W9-CHJ-111

Early NASA Manned Space Flights

Spacecraft	Crew	Launch Date	Duration	Remarks
Project Mercury				
Mercury-Redstone 3	Alan Shepard	May 5, 1961	15 minutes, 22 seconds	First American in space (suborbital).
Mercury-Redstone 4	Virgil "Gus" Grissom	July 21, 1961	15 minutes, 37 seconds	Second suborbital flight; spacecraft sank, Grissom rescued.
Mercury-Atlas 6	John Glenn	February 20, 1962	4 hours, 55 minutes	First American in orbit (3 orbits); telemetry falsely indicated heat-shield unlatched.
Mercury-Atlas 7	M. Scott Carpenter	May 24, 1962	4 hours, 56 minutes	Retrofire error caused 250-mile landing overshoot.
Mercury-Atlas 8	Walter Schirra	October 3, 1962	9 hours, 13 minutes	Orbited Earth six times; closest splashdown to target to date (4.5 miles).
Mercury-Atlas 9	Gordon Cooper	May 15, 1963	1 day, 10 hours, 20 minutes	First U.S. mission to last longer than a day; made 22 orbits; manual re-entry necessitated by systems failure; landed 4 miles from target.
Project Gemini				
Gemini 3	Virgil "Gus" Grissom, John Young	March 23, 1965	4 hours, 53 minutes	First two-person U.S. crew; first piloted spacecraft to change orbital path; first onboard computer.
Gemini IV	James McDivitt, Edward White	June 3, 1965	4 days, 1 hour, 56 minutes	First U.S. EVA (White, 21 minutes); first U.S. four-day mission; computer failure necessitated manual re-entry.
Gemini V	Gordon Cooper, Charles Conrad	August 21, 1965	7 days, 22 hours, 56 minutes	First use of fuel cells for electric power.
Gemini VII	Frank Borman, James Lovell	December 4, 1965	13 days, 18 hours, 35 minutes	Longest mission up until that time; orbited Earth 206 times; first U.S. simultaneous missions (with *Gemini VIa*).
Gemini VIa	Walter Schirra, Thomas Stafford	December 15, 1965	1 day, 1 hour, 51 minutes	Came to within six feet of *Gemini VII*.
Gemini VIII	Neil Armstrong, David Scott	March 16, 1966	10 hours, 41 minutes	First docking in space (with an Agena target vehicle); control malfunction necessitated emergency re-entry; first splashdown in the Pacific Ocean.
Gemini IX	Thomas Stafford, Eugene Cernan	June 3, 1966	3 days, 21 minutes	EVA (2 hours, 1 minute, Cernan); splashdown only one-half mile from target ship.
Gemini X	John Young, Michael Collins	July 18, 1966	2 days, 22 hours, 47 minutes	Docked with Agena target vehicle; attained record altitude (474 miles); EVA (39 minutes, Collins).

ALPHA

tear here

Spacecraft	Crew	Launch Date	Duration	Remarks
Gemini XI	Charles Conrad, Richard Gordon	September 12, 1966	2 days, 23 hours, 17 minutes	Using an Agena engine, set altitude record (850 miles); EVA (2 hours, 43 minutes, Gordon, connected *Gemini* and Agena by tether); first computer-guided re-entry.
Gemini XII	James Lovell, Edwin Aldrin	November 11, 1966	3 days, 22 hours, 34 minutes	EVA (5 hours, 30 minutes, Aldrin); docked with Agena; last Gemini mission.

Apollo Program

Spacecraft	Crew	Launch Date	Duration	Remarks
Apollo 7	Walter Schirra, Donn Eisele, Walter Cunningham	October 11, 1968	10 days, 20 hours, 9 minutes	First manned Apollo mission; first U.S. mission with a three-person crew; first live TV from space.
Apollo 8	Frank Borman, James Lovell, William Anders	December 21, 1968	6 days, 3 hours	First manned mission to go into lunar orbit; first use of a Saturn V rocket for a manned mission; live TV of lunar surface.
Apollo 9	James McDivitt, David Scott, Russell Schweickart	March 3, 1969	10 days, 1 hour, 1 minute	Schweickart tests lunar space suit during EVA in Earth orbit; first piloted flight of lunar module.
Apollo 10	Thomas Strafford, John Young, Eugene Cernan	May 18, 1969	8 days, 3 minutes	First lunar orbit in lunar module; came within 50,000 feet of Earth's surface; set manned speed record, 6.8863 miles per second during re-entry.
Apollo 11	Neil Armstrong, Michael Collins, Edwin Aldrin	July 16, 1969	8 days, 3 hours, 18 minutes	First lunar landing; Armstrong first man on Moon; Armstrong and Aldrin 151 minutes EVA on lunar surface, collected 48½ lbs. of Moon rocks; time on Moon: 21 hours, 36 minutes.
Apollo 12	Charles Conrad, Richard Gordon, Alan Bean	November 14, 1969	10 days, 4 hours, 36 minutes	Second manned Moon landing; Conrad and Bean on Moon for 31 hours, 31 minutes.
Apollo 13	James Lovell, John Swigart, Fred Haise	April 11, 1970	5 days, 22 hours, 55 minutes	Service module oxygen tank exploded, causing mission to be aborted; crew returned safely to Earth using lunar module; mission holds the world altitude record (248,665 miles).
Apollo 14	Alan Shepard, Stuart Roosa, Edgar Mitchell	January 31, 1971	9 days, 42 minutes	Third manned Moon landing.
Apollo 15	David Scott, Alfred Worden, James Irwin	July 26, 1971	12 days, 7 hours, 12 minutes	Scott and Irwin made fourth Moon landing; first lunar rover use; first deep space walk.
Apollo 16	John Young, Thomas Mattingly, Charles Duke	April 16, 1972	11 days, 1 hour, 51 minutes	Fifth manned Moon landing.
Apollo 17	Eugene Cernan, Ronald Evans, Harrison Schmitt	December 7, 1972	12 days, 13 hours, 51 minutes	Sixth, and last, manned lunar landing.

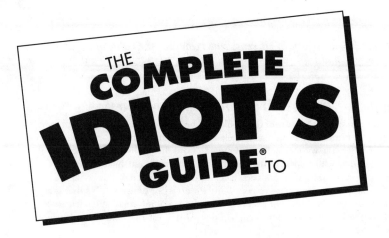

THE

COMPLETE IDIOT'S GUIDE® TO

NASA

*by Veteran Astronaut Thomas D. Jones, Ph.D.
and Michael Benson*

ALPHA

A Pearson Education Company

To my fellow American citizens, who recognize the value of investing in the future, and whose support of space exploration will keep our nation on the vital frontier.

Publisher: *Marie Butler-Knight*
Product Manager: *Phil Kitchel*
Managing Editor: *Jennifer Chisholm*
Acquisitions Editor: *Gary Goldstein*
Development Editor: *Michael Thomas*
Production Editor: *Katherin Bidwell*
Copy Editor: *Cari Luna*
Illustrator: *Chris Eliopoulos*
Cover/Book Designer: *Trina Wurst*
Indexer: *Ginny Bess*
Layout/Proofreading: *Angela Calvert, Brad Lenser, and John Etchison*

Contents at a Glance

Contents

Foreword

Within this book Michael Benson and Tom Jones present a history of spaceflight—suborbital, orbital, satellite, robotic, and human—that is readable and understandable on one level and intellectual and romantic on another. This book is both big and beautiful; it has a monumental scope but focuses on the essentials. It answers not only the what but the why, and provides a sense of the past, the present, and our possible trajectory into the future.

Michael is an experienced and talented writer; Tom is a renaissance man, a wonderful human being, and an astronaut completely at home in space. Michael weaves an intricate pattern from the materials of spaceflight history; Tom colors and textures the tapestry from his experience and emotions, having been there physically, scientifically, and spiritually. Together they present the objective facts of spaceflight history in a style that is pleasant, playful, personal, and reminiscent of the romantics.

I worked and trained with Tom for years, and flew with him for eighteen days during mission STS-80 in November of 1996. I have flown in space with 26 different people; Tom is as completely immersed in the space environment as anyone I have ever flown with. He is extraordinarily talented in a technical sense but equally so in his ability to relate the experience of spaceflight. He is sensitive, open, and creative. As a former B-52 commander he possesses a wonderful and powerful empathy for his fellow crewmembers, which enables him to express the sensations and perceptions of spaceflight to others. As a planetary scientist he not only knows the early history of planetary probes but he can express it as if he were there, riding along through the solar system. At work as a scientist in orbit he knows—and just as important, feels—the heavens and the earth; they are personal to him. As a great and strong human being, he has the courage to surrender to the universe and allow it to do with him what it will.

This book is a well-researched and -documented history of spaceflight, but it also reflects an understanding that can only be acquired by giving one's self over to the space environment—the free-fall, the heavens, and the Earth as experienced from a unique vantage point. It is a history—a history seen and felt with a sense of awe and wonder—but also a story of the future, the potential, the possible.

—Story Musgrave, M.D.
NASA Astronaut, 1967–1997

Introduction

Ground Control to Dr. Tom: About Mission Specialist Thomas D. Jones, Ph.D.

Before we start learning about NASA, let me begin by introducing myself. I'm Thomas D. Jones, former astronaut. Pleased to meet you!

I grew up in Baltimore, Maryland, in Essex, just outside the city, on Chesapeake Bay. My earliest memories of Essex include airplanes flying over our town from the big Martin aircraft factory about three miles away. So airplanes were in my blood early. In school, I sketched them, traced them, day-dreamed about them—flying was the ultimate dream.

Then in the 1960s, the excitement of the Moon race hooked me, and I saw how airplanes could lead me to spacecraft that traveled beyond the atmosphere.

The astronauts of the 1960s were all test pilots, and if I were to combine my twin loves of science and aviation and eventually fly in space, I'd have to become a test pilot, too.

So it was off to the Air Force Academy in Colorado Springs, where I majored in Basic Sciences and learned how to fly at last. Wearing my shiny new, gold second lieutenant's bars, I headed for pilot training.

Though I longed to be a fighter pilot, the Air Force in its infinite wisdom sent me to a bomber squadron, and I was soon a co-pilot on a B-52 Stratofortress. Moving up to aircraft commander, I spent five years with my crews practicing for a nuclear delivery mission that—blessedly—never came. With the Air Force needing few bomber test pilots, I resigned as a captain in 1983 to pursue my long-held interest in space science.

At the University of Arizona in Tucson, I went back to the books in planetary science, the study of the planets in our solar system and the processes that shape them. From a telescope atop a dormant Hawaiian volcano, I surveyed about 40 asteroids that orbit between Mars and Jupiter, searching for water in their surface minerals. Matching those data with meteorites in our lab, my colleagues and I learned that a majority of the darkest main belt asteroids still hold water in their rocks, making them some of the least-altered—and resource-rich—objects in the solar system.

In 1988, with my doctorate in hand, I put my technical skills to work on remote sensing problems for the CIA, the Central Intelligence Agency.

A year later I was back in planetary science, helping NASA plan future robotic missions to Mars, the asteroids, and the outer planets.

I'd sent in an astronaut application while with the CIA, but didn't have much hope for getting hired—I'd already been turned down twice! So I was floored when NASA invited me to interview for a mission specialist astronaut's job in the fall of 1989.

Against all odds—NASA always has at least 2,000 qualified applicants on file—I managed to make the cut and join the Astronaut Class of 1990. After a year of "basic training," I became eligible for spaceflight in 1991.

Here I am during a launch pad rehearsal for my final mission in space in January 2001. That's me with the grin.

Since that exhilarating start 11 years ago, I've had the privilege of flying four times in space, on three of the four shuttles: *Endeavour*, *Columbia*, and *Atlantis*. I've logged about 53 days circling the planet, with the highlight being my three spacewalks outside the International Space Station in February 2001. Nineteen hours outside in a space suit—the experience has been beyond the wildest dreams of my youth.

Now I'm a retired astronaut, and it is my pleasure to write this beginner's guide to the National Aeronautics and Space Administration, better known as NASA. It is my hope that this book will pique your interest enough to read further about NASA and space exploration.

This book is designed provide you with a brief history of NASA, explain what NASA is all about today, offer advice and encouragement to those who might want to work for NASA someday—as an astronaut or otherwise—and offer tips to those who are considering visiting one of NASA's sites while on vacation.

Here is an idea of what you can expect in the pages to come:

Part 1: "From Tang, to the Moon, and Now the Universe." Part 1 covers my four missions in space, a review of why space exploration is so important, an introductory look at physics and robotics to give you a general idea of how things work, tips for prospective astronauts (and others considering a career with NASA), and tips for tourists considering a visit to one of NASA's sites.

Part 2: "The Early Days." We meet the founding fathers of rocketry and follow the development of rocket science up to the successful launch of the first satellites to orbit the Earth. We'll also take a look at early weather and communications satellites and how they changed the world.

Part 3: "Humans in Space." The first people in space were true pioneers, and this is their story, from Yuri Gagarin's first orbit around the Earth to the Project Gemini missions that prepared us for our first trip to the Moon.

Part 4: "Moonwalkin'." These are the missions in which men visited the Moon. This section includes the story of *Apollo 13*—the "Houston, we have a problem" mission.

Part 5: "After the Moon." Since the end of the Apollo program, NASA has not only used cameras and other equipment on satellites to look inward and learn things about our planet, they have turned those instruments outward as well to give us new clues to the origins and fate of our universe.

Extras

On every page, interspersed among the text, you will find a plethora of sidebars. These sidebars include:

Space Talk
In "Space Talk" sidebars, I define words and terms that might be unfamiliar to you.

Cosmic Facts
"Cosmic Facts" sidebars are self-explanatory—you'll find facts and trivia about space and space travel here.

Dr. Jones's Corner
In "Dr. Jones's Corner," I'll offer personal observations on various subjects.

Astro Bio
You'll also find "Astro Bio" sidebars, which contain short biographies of Mercury, Gemini, and Apollo astronauts and other spaceflight figures.

Acknowledgments

I'm able to share the story of spaceflight with you thanks to the contributions of many friends and colleagues. My parents gave support and encouragement to a 10-year-old's dream of flying in space one day. My teachers gave me the discipline to learn, a willingness to study, and a thirst for knowledge. My wife and children stood by me through years of schooling, training, and four nerve-racking space shuttle missions. My instructors gave me the skills that kept me alive in orbit and allowed me to accomplish our mission. My flight controllers and fellow astronauts watched over me and brought me into the most exclusive team of professionals working today—I've been honored to be a part of your efforts. You keep the dream of space exploration vital and exciting as you successfully turn our dreams into reality.

Special thanks also go to Glen Golightly at The Boeing Company; editor Gary Goldstein; Mike Gentry and Eileen Hawley at the Johnson Space Center; Margaret Persinger and Patti Beck at the Kennedy Space Center; and Jack King of the United Space Alliance.

For their contributions to the writing of this book, I'm especially grateful for the encouragement of Dale Brown, the trust of our agent Jake Elwell, the confidence of our editor Michael Thomas, and the amazing energy of my co-author, Michael Benson. His professionalism, thoroughness, and pursuit of accuracy make me suspect he's been a NASA explorer himself.

Trademarks

All terms mentioned in this book that are known to be or are suspected of being trademarks or service marks have been appropriately capitalized. Alpha Books and Pearson Education, Inc., cannot attest to the accuracy of this information. Use of a term in this book should not be regarded as affecting the validity of any trademark or service mark.

Part 1

From Tang, to the Moon, and Now the Universe

I've been a fan of NASA since I was a boy, and working with the space agency as an astronaut was the high point of my career. So forgive my enthusiasm for its story—it's a great one to tell! In Part 1, I'm going to kick things off by telling you about what I know best—my own experiences in space. It's a privilege to represent our country in space and an honor to circle the Earth with some of the most talented people on—or off—the globe.

Then we'll look at NASA as it exists today, how it works and what its goals are. We'll learn how rockets work and how we get our satellites—and shuttles—off the planet. I'll offer tips for those who dream of being an astronaut—or perhaps working for NASA in another capacity. We'll close this section with information and suggestions about visiting NASA's facilities.

As Konstantin Tsiolkovsky (you'll learn about him in Chapter 3) said: "Earth is the cradle of mankind, but one cannot stay in the cradle forever."

My Adventures in Space

In This Chapter

♦ Mapping the Earth with radar

♦ Eight-and-a-half very long minutes

♦ Volcanic video

♦ Searches for the origins of the universe

As a space shuttle astronaut, I first went into space on April 9, 1994, aboard the Space Shuttle *Endeavour*. The shuttle blasted off from Pad 39A at the Kennedy Space Center at 7:05 in the morning. The launch had been originally scheduled for April 7, but was delayed so that the liquid-oxygen preburner (don't worry, you don't need to know what that is) could be inspected. The launch was rescheduled for April 8, but again had to be postponed, this time because of a low cloud cover and a strong crosswind. My crewmates were commander Sidney M. Gutierrez, pilot Kevin P. Chilton, payload commander Linda M. Godwin, and mission specialists Jay Apt and Michael R. Clifford.

First Flight of the Space Radar Laboratory

After a series of space shuttle missions geared toward improving our view outward into the universe, this mission turned our attention around and focused instruments on the Earth. As soon as we got into space we began to set up the equipment that would take the pictures. And these were not just any old cameras.

Ready to go! My official astronaut portrait.

(NASA)

Dr. Jones's Corner

During my first mission in space, *Endeavour* orbited the Earth 183 times at a height of 121 nautical miles. (A nautical mile is 6,076.1 feet, or 1.15 miles.) The mission lasted 11 days, 5 hours, 49 minutes, 30 seconds, and we traveled 4,704,875 miles.

The system NASA constructed was called Space Radar Laboratory (SRL). The Space Radar Laboratory (SRL) payload was composed of the Spaceborne Imaging Radar-C/X-Band Synthetic Aperture Radar (SIR-C/X-SAR) and the Measurement of Air Pollution from Satellite (MAPS). This versatile new payload was built by NASA and its partners, the German and Italian space agencies. It was designed to gather information that would enable us to track changes in the environment and distinguish between natural and human-caused effects.

Our imaging radar could measure the Earth's surface regardless of weather conditions. It could see through clouds and didn't need sunlight, so we could peer down at Earth around the clock, from the Aleutians to the lower-most tip of South America. The radar waves penetrate clouds, and under certain conditions, can also "see" through vegetation, ice, and extremely dry sand. In many cases, radar from above is the only efficient way scientists can explore inaccessible regions of the Earth's surface.

On April 11, I was able to give scientists real-time observations of thunderstorms over Taiwan, the Philippines, and New Guinea to augment data being gathered by the MAPS experiment.

After scanning the Earth for 11 days, the data recorded during the *STS*-59 mission (STS stands for Space Transportation System) would fill the equivalent of 20,000 encyclopedia volumes. More than 27 million square miles of the Earth's surface, including land and sea, were mapped on this flight, 12 percent of Earth's total surface. The Space Radar Laboratory (SRL) obtained radar images of about one quarter of the planet's land surfaces.

Cosmic Facts

Top Five Space Recreations

- Looking out the window at Earth
- Space "gymnastics"
- Photography
- Exercise
- Reading and listening to music

Back to Earth

After waiting a day for rainy weather at the Cape to clear, we came back down to Earth at Edwards Air Force Base in California on April 20, 1994. Both *Endeavour* and SRL were immediately turned around by the NASA team to fly again.

That was not the last time that bad weather on Earth kept me in space for longer than I had planned to be—not by a long shot! In spaceflight, the unexpected is commonplace.

From Earth to Space: Eight-and-a-Half Very Long Minutes

When riding the space shuttle, it takes eight and a half minutes to get from lift-off to space, from the Earth to a weightless condition. I doubt if there are eight-and-a-half longer and more thrilling minutes in anyone's life, especially the first time one does it!

The just-launched space shuttle passes the speed of sound, going nearly straight up, in about 45 seconds. Talk about blowing town! We accelerate so quickly that by the time we pass the top of the launch tower, we're already going 100 miles per hour.

Cosmic Facts

Outside the Earth's atmosphere and in orbit around the planet, astronauts experience a state of free-fall, or weightlessness. Inside a spacecraft, because everything is falling, it appears that nothing falls. There is no up or down. That's why we say things float in space. Free-fall can't be distinguished from "floating."

When the orbiter returns to Earth and approaches the Cape still going faster than Mach 1, ground observers hear two *sonic booms*, one caused by the spacecraft's nose and the other by its tail. There are two sonic booms after lift-off as well, but they can't be heard by ground observers because of the roar of the engines.

STS-68

The SRL and I returned to space on STS-68, once again aboard *Endeavour*. That mission began at 7:16 in the morning on September 30, 1994—although there were times when I wondered if this mission was ever going to get underway.

Space Talk

A **sonic boom,** which resembles the cracking of nearby thunder, is caused when an object moving faster than the speed of sound generates a shock wave as the vehicle forces its way through the air. The speed of sound is also known as Mach 1. To attain Mach 2, one must travel twice the speed of sound.

The mission had been scheduled to begin on August 18, 1994, and the countdown had gone smoothly. (The countdown is there so that all systems can be checked and double-checked at precisely the correct moment. The countdown can be halted if there is a problem.) In fact, the countdown got all the way to within one second of zero, and the main engines had ignited. Those engines burned for almost two seconds before they were automatically shut off—our onboard computers detected overheating in one of the engines. We shut down safely, just as the system was designed to do, but the problem ended up causing an almost six-week delay.

But on September 30, we finally made it into space and once again we turned our radar imagers toward the Earth. My crewmates this time were commander Michael A. Baker, pilot Terrence W. Wilcutt, and mission specialists Steven L. Smith, Daniel W. Bursch, and Peter J.K. Wisoff.

The delay had proved lucky. One of the highlights came when, over the course of a week, we studied in detail the Kliuchevskoi (pronounced *clue-chev-skoy*) volcano erupting in Kamchatka.

My second spaceflight concluded at 1:02 on the afternoon of October 11, 1994, as *Endeavour* touched down on Runway 22 at Edwards Air Force Base. Again, bad weather forced us to land in California, but we at least got an extra day in orbit.

Endeavour had orbited the Earth at an altitude of 120 nautical miles. During the mission, we circled the Earth 182 times, traveling 4.7 million miles. The satisfying mission lasted 11 days, 5 hours, 46 minutes, and 8 seconds. But there was so much more to do in space.

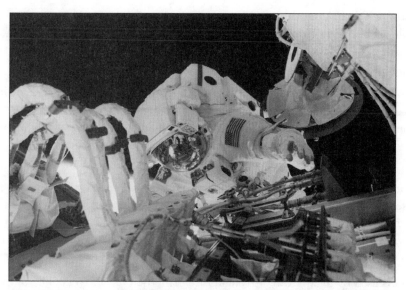

Here I am working outside the International Space Station (ISS) on my first space walk, Feb. 10, 2001. Crewmate Bob Curbeam and I connected these vital electrical and fluid lines between the ISS and the new Destiny laboratory, which we installed on the STS-98 mission, February 2001.

(Photo by space walker Bob Curbeam and NASA)

Anatomy of My Space Suit

My EVA suit (or EMU—Extravehicular Mobility Unit) is modular—that is, it has several interchangeable components. My space suit was designed to protect me from the hot and cold temperatures, as well as the vacuum, of space. Here are the major parts that make up a space suit, and a quick explanation of how each helps protect the astronaut:

The helmet. The helmet channels cool, fresh oxygen to the astronaut's face. The clear, double-pane helmet bubble is made of high-impact Lexan, a kind of plastic. A plastic and metal cowl over the helmet protects it from impacts, and mounts both work lights and TV cameras.

The Hard Upper Torso, or HUT. The top portion of the suit, called the HUT, is made of fiberglass and steel and provides a hard shell of protection around the astronaut's chest, back, and shoulders. Flexible metal bearings at the shoulders permit the arms to rotate and swivel slightly, front to back. The arms and gloves are adjustable in length and attach to the shoulder bearings. The helmet attaches to a metal seal at the top of the HUT, which is built right into the life-support backpack.

The Lower Torso Assembly, or LTA. The "pants" of the suit. The waist area and legs are made of soft layers of insulation, neoprene rubber, and kevlar, which both hold pressure and protect the astronaut. The LTA bends at a waist ring and at knee joints and has attach points at the bottom for the boots.

The Portable Life Support System backpack, or PLSS. The PLSS contains the machinery needed to keep an astronaut alive in the harsh environment of space. This pack, built into the rear of the hard upper torso, contains the suit's oxygen, battery, and water supplies. The PLSS also has dual radios to provide communications back to the orbiter, the space station, and Mission Control.

The Liquid Cooling and Ventilation Garment, or LCVG. The LCVG is the astronaut's set of "long underwear." The spandex body suit is laced with spaghetti-thin water lines, which circulate cooled water from the backpack next to the astronaut's skin. The warm water and air collected by the tubes and by finger-thick ventilation ducts is piped back to the backpack for cooling and removal of carbon dioxide.

Space suits used by today's U.S. astronauts are composed of 12 layers. The outside of the space suit is made of a combination of Gortex, Kevlar, and Nomex. Kevlar is the material used to make bullet-proof vests and it is there to protect astronauts from micrometeorites, which are much like tiny bullets shooting through space. The seven layers inside of that are all designed to protect the astronaut from the extreme temperatures of space, and they are made of aluminized Mylar, laminated with Dacron scrim. Inside of that is a layer designed to maintain the proper pressure inside the suit. It is called the pressure-bladder layer and is made of urethane-coated nylon surrounded by a layer of pressure-restraining Dacron. Peel that away and you find the inner-most layer, which is the cooling system made of spandex with tubing sewn in, called the LCVG.

Although the space suit weighs 280 pounds on Earth, it weighs nothing in space. An astronaut has supplies in his space suit to work outside for up to eight hours.

My Third Mission: STS-80

My third mission in space was designated STS-80. My crewmates were Kenneth D. Cockrell, mission commander; Kent V. Rominger, pilot; and mission specialists Tamara E. Jernigan and F. Story Musgrave. We left Earth aboard *Columbia* on November 19, 1996, and stayed aloft for a record-breaking shuttle flight lasting 17 days, 15 hours, and 53 minutes—not that I was counting.

All That the Shuttle Was Built to Do

STS-80 showcased all the capabilities that the space shuttle had acquired over 15 years of flying. On our mission, *Columbia* launched and retrieved two different satellites for scientific research into astronomy and the commercial use of space. Two rendezvous were accomplished and executed in an efficient manner so that a minimum of fuel was used up. (That was great practice for a future space station link-up.)

No Ifs, Ands, *ORFEUS-SPAS*

What the heck is *ORFEUS-SPAS*, you ask? That's the name of the first satellite we launched on this mission. It was a telescope that sees a wavelength of light that is not able to fully penetrate the Earth's atmosphere—*ultraviolet* light.

A little bit of ultraviolet light comes down through the atmosphere and gives us sunburns. Those are the same rays that the ozone layer filters out as sunlight streams down through the atmosphere. We filter out most ultraviolet light with ozone.

We want to study ultraviolet light because it is emitted from very hot objects in the galaxy and the universe. The centers of *galaxies*, *black holes*, *white dwarfs*, the remnants of *supernovae*, and similar kinds of hot sources emit information about their composition and activity in the ultraviolet portion of the spectrum. Since that light can't make its way down to observatories on the Earth, we sent our telescope up above the atmosphere.

Space Talk _____

Ultraviolet light is light that has a wavelength so short that the human eye cannot see it.

Cosmic Facts _____

If you totally relax all your muscles in space, you will take on a posture a bit like that of an ape or an unborn baby. Your shoulders will hunch slightly and your arms and legs will come up in front of your body. Float in a pool and relax all your muscles, and you'll find that the same thing happens!

Dr. Jones's Corner

When we say the space shuttle *launched* a satellite, we mean that we took the satellite up into space with us on the space shuttle and then pushed it out once we were in orbit, so that the satellite would go off to do its work on its own. When I say we *retrieved* a satellite, I mean we rendezvoused with it, grabbed it using our shuttle's robotic arm, and pulled it back into the space shuttle. When we came back to Earth, we took the satellite—and the information it contained—back down with us.

We can now look across the universe, across our galaxy and learn more about the physics of these very energetic objects. Even though they're in our stellar neighborhood, they date way back to the origins of the universe. They are of great interest in terms of understanding the physics of our universe.

The Hubble Space Telescope, which I talk more about in Chapter 29, "Revealing the Planets: Today's Robotic Explorers," can also use the ultraviolet spectrum to study the same objects. But *ORFEUS-SPAS* is a telescope designed to go even farther into the more energetic region of the spectrum and look at the ultraviolet light that the Hubble Telescope would miss. So we had a new capability to look at these energetic objects. That was why we left the satellite in orbit for 14 days before retrieving it. That way we got the maximum amount of information about the physics of our violent universe.

IMAPS

IMAPS, made at Princeton University in New Jersey, was one of the two telescopes on *ORFEUS-SPAS*. *ORFEUS-SPAS* has a big one-meter mirror that gathered ultraviolet light and directed it to two spectrographs attached to that central telescope. IMAPS was a smaller scope, with a very high-resolution spectrograph. A spectrograph splits the light into very fine detailed lines so that we could see specific atoms that were emitting or absorbing this light. So it gave us more information about the specific energy state and composition of an object.

IMAPS couldn't see as far off into the universe as the more sensitive main telescope, but it gave us more information about the composition and the temperatures and the details of the physics. Cooperating with NASA regarding the satellite's main telescope were the University of Tubingen in Germany and the University of California at Berkeley.

Cosmic Facts

It's easy to move about in free-fall. Actually, it's too easy. When the slightest push sends you flying across the space shuttle, you learn to be careful with the way you move your body. When we work in the space shuttle, the challenge is to stay still, not to move. Without a restraint system we would float away from the place where we were working, and that could get to be pretty annoying. The system we use is foot restraints. Hook your feet into these heavy cloth loops and you don't float away. Space shuttle astronauts have learned to move around by pushing off walls and other fixed objects with their fingertips.

Some Like It Hot

The importance of learning about subjects as esoteric as black holes or white dwarfs that are light years away from us is not so much to learn about our own local environment in the solar system, but more to teach us about the laws of physics—and the eventual fate of our Sun.

The laws of physics that you see in these highly energetic objects—that are very hot, much hotter than the surface of the stars in our vicinity like the Sun and its local neighbors—give us clues about the formation of our universe. So these high temperatures that we now only see in the left-overs of a supernova, or the swirling disk around a black hole, or the energetic, hugely powerful core of a galaxy that's a billion *light years* away mimic the very high temperatures that occurred just after the *big bang*.

If we can see how the laws of physics operate today, in these small local environments that are very difficult to observe from the ground, we have found a way to discover how the rules operated back at the beginning of the universe and if the rules have changed since then.

In addition to deploying and retrieving both *ORFEUS-SPAS* and the *Wakeshield Facility* (more about that in a second), we were scheduled to accomplish two space walks to test the adequacy of our tools and techniques for the construction of the International Space Station (ISS). However, to our extreme disappointment, Tammy Jernigan and I had our space walks cancelled because of a jammed external hatch on *Columbia*. Rather than force the hatch open, and risk having problems closing it later (!), Mission Control advised the safe route and we gave up on the space walks. NASA rescheduled them, and our friends on another crew accomplished the work a year later.

The other objective of the flight was to use our planned 16 days in orbit to conduct a variety of materials science and biology investigations in the mid-deck of the shuttle.

Efficiency was the key. Even as *Columbia* was maneuvering as a versatile satellite launcher, it was also serving as a science laboratory.

Space Talk

A **light year** is the distance that light travels in a year. With light traveling at 186,000 miles per second, that's about six trillion miles.

The **big bang** theory states that the universe was born with one gigantic explosion, thus explaining why everything is in motion and the universe is growing.

Cosmic Facts

Short-Distance Phone Calls Department: Even if two space-walking astronauts were working side by side outside the space shuttle, they would have to use their radios to talk to one another. If they tried to speak to one another without electronic help, their voices would not carry, because there is nothing to vibrate and transmit sound in the vacuum of space.

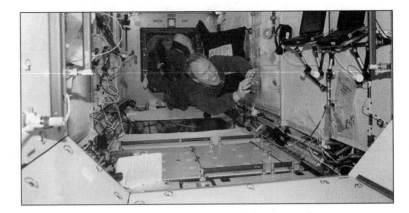

Inside the new Destiny laboratory, Mission Commander Ken Cockrell and I drift in free-fall down the lab's center aisle. The U.S.-built lab, the centerpiece of the International Space Station, is 28 feet long and 14 feet wide, and weighs about 32,000 pounds.

(NASA)

Wakeshield Facility

Our second satellite on STS-80 was the *Wakeshield Facility*, a mini-factory for making more efficient computer chips—in space! The WSF was built by NASA and the University of Houston, and it tested a technique for making ultra-pure, more efficient semi-conductor wafers in the high vacuum of space. These wafers, produced under contamination-free conditions, might outpace ground-produced chips and lead to faster, cheaper, and more competitive computer chips for the electronics industry.

My job on STS-80 was to release *Wakeshield* into orbit using *Columbia*'s robot manipulator arm. On the fourth day of the flight, I grappled *Wakeshield*, and after a long checkout procedure while we held the satellite just outside the payload bay, we released it into orbit. As *Wakeshield* slowly moved off, it coasted just ten feet above our cockpit windows. We had a superb view, and in our nervousness we joked that we ought to have designed *Wakeshield* with curb feelers! The satellite took up its station about 30 miles behind *Columbia*, and the same distance in front of *ORFEUS-SPAS*.

SVS

The Space Vision System, or SVS, is one of the systems we experimented with during our satellite retrievals. The SVS is a way of processing high-quality television pictures to determine where an object is in three dimensions. An astronaut can then use depth perception (looking out the window), TV monitors, and the computer displays available to him in the cockpit to judge where his satellite target is and move accordingly. The SVS has since played a prime role in assembling modules to the space station in situations where both TV and out-the-window views are blocked by station hardware. It's an efficient way to make sure that a billion-dollar module is going just where it's supposed to.

Wanna Arm Wrestle?: Using a Robot Arm in Orbit

Three days later, its chip-manufacturing work complete, commander Ken Cockrell pulled *Columbia* up beneath *Wakeshield*, hovering about 15 feet below the satellite. My hands on the manipulator joysticks, my eyes switching between the *Wakeshield* looming outside the windows and the TV monitors on the flight deck, I moved in with the arm for the grapple.

The arm handles very smoothly in orbit, but there was a little oscillation side-to-side as I moved in over the grapple pin. I'm not sure what was shaking more—my hands on the controls, or the arm's end effector, swaying ever-so-slightly because of my initial control input. But a moment later I had nestled the end effector—it looks like a big soda can—over the *Wakeshield*'s grapple pin, and snapped closed the wire snares that snug the satellite up to the arm. One aboard, one to go!

With Tammy Jernigan at the arm controls, *ORFEUS-SPAS* came smoothly aboard on Flight Day 15. This mission marked the first time that a shuttle had released and retrieved two satellites, and flown formation with both in Earth orbit. Quite a navigational and robotics triumph for our versatile shuttle.

STS-98

My fourth—and most exciting—journey into space was aboard the Space Shuttle *Atlantis*. On board with me were commander Kenneth D. Cockrell, pilot Mark L. Polansky, and mission specialists Robert L. Curbeam and Marsha S. Ivins. We lifted off from the Kennedy Space Center on February 7, 2001, at 6:13 P.M. Eastern Standard Time. It was a beautiful sunset lift-off. The mission was designated STS-98, the seventh shuttle mission designed to assemble Alpha, the International Space Station (ISS). (For more about the ISS, see Chapter 27, "The International Space Station—Call Sign 'Alpha'.") The piece we had to install was the $1.4 billion U.S. laboratory module, Destiny. We spent nearly seven days docked to the station while we attached and activated the new lab.

Our mission to the International Space Station turned out to be a milestone in NASA's history. The installation of the Destiny science lab made space station Alpha officially history's biggest space habitat. More important, Destiny took over prime control of the station's computer, life support, and power distribution systems, enabling the start of scientific research aboard the ISS.

Three Space Walks

The highlight of the mission for me—and the biggest challenge—was the set of three space walks that were essential to activating and equipping Destiny with its power, cooling, and computer lifelines. After being skunked on STS-80 by our balky hatch, I was taking nothing for granted. Through the more than 200 hours that my colleagues Mark Lee and Bob Curbeam and I spent underwater, training for these complex EVAs, I tried to keep myself from anticipating the exhilaration of a space walk too much. I wouldn't believe it was happening until I emerged from *Atlantis*'s airlock hatch for the first time.

Our first EVA was the most important. While Bob and I readied for the space walk on Flight Day 4, with Mark Polansky as our in-cabin choreographer, Ken Cockrell and Marsha Ivins worked to get Destiny up onto the space station. My first job out of the hatch—this space walk was for real!—was to help Marsha move a docking port temporarily out of the way, making way for the new Lab. Marsha swung the Lab ever-so-carefully out of the payload bay, then swapped Destiny end-for-end to line up its docking port with the space station's. She did the job perfectly, and Ken ("Taco" to his friends) drove the remote-controlled bolts that locked Destiny permanently to the ISS.

Here I am on the second of my three space walks during the STS-98 mission, waving to my crewmates inside Atlantis's flight deck. I'm holding on to a handrail on the International Space Station's Unity node; the solar array wings atop the station, stretching 240 feet from tip to tip, extend behind me. I'm carrying an array of tools on the chest of my spacesuit, and a portable foot restraint is tethered by my left hip.

(Photo by crewmate Marsha Ivins and NASA)

Now it was our turn. Bob and I teamed up to attach critical electrical power and cooling lines to Destiny from the station. We floated gingerly around the new Lab, mating stiff, bulky connectors, and stringing cable for a backup heater system. The crucial moments came as we prepared to hook up four ammonia cooling lines to Destiny, providing it with its vital heat-removal capability. Without these lines in place, the Lab's systems would overheat and fail soon after they were activated.

Bob had the difficult task of bending these thick hoses over to Destiny. As he released the first line from the station, he was startled by a jet of frozen ammonia crystals, spurting from a faulty valve like the nozzle of a snow-maker. The leak could deplete the ammonia cooling supply if not isolated, so he moved quickly to remove the adjacent hose, feeding ammonia to the leak. As I moved up the Lab's forward end to help, I could see the cloud of vapor and ice crystals expanding rapidly into space, streaming out like a comet tail. But Bob was on top of the situation. He popped loose the adjoining hose, cutting off the supply of ammonia to the leak. Only that first valve proved troublesome, and the other three connections went smoothly. Seven and a half hours after floating outside, we had all the cooling and electrical connections completed. Destiny was in business.

Delayed Landing

We had two more space walks to outfit the Lab and station for future assembly missions, but our work went very smoothly. Inside the station, we activated the Lab with the help of the first Expedition Crew, who'd been aboard for the past three months. After transferring spare parts, scientific gear, and supplies, it was finally time to say good-bye. We undocked from ISS, leaving the new Destiny module sparkling like a gem at the front of the station. We had a welcome break in orbit over the next day. We had time for a workout on the exercise bike, and a chance finally to look at our planet and bring back some photography for the Earth scientists back home.

In the end we wound up with a couple of bonus days in orbit, bringing our flight duration to nearly 13 days. Once again, as on my first and second missions, the Florida weather seemed determined to keep us from our Kennedy Space Center runway. With our supplies running low, Mission Control bit the bullet and decided to bring us in to Edwards Air Force Base. With Taco swinging *Atlantis* in over Runway 22, we touched down gently just after noon in the California high desert, our crew exuberant over the success of our mission. It'd be impossible to top that last trip in orbit!

Dr. Jones's Corner

To keep our hands free for work while outside on a space walk, the shuttle and station carry moveable "foot restraints," similar to stirrups, that we set up at our various work sites. The foot restraints allow us to slip our boots into place, which both hold us in place and leave our hands free to deal with tools and equipment.

Oh, All Right, Here's How We Go to the Bathroom

Space is still a tough environment for bathroom technology. The shuttle has a practical space toilet that has nearly all the bugs worked out. Instead of using gravity to make things go where they're supposed to, our cramped, airliner-style "waste control system" (WCS) compartment uses moving air.

Urination can be done while simply floating in front of the WCS compartment. Using an electric fan, the WCS sucks urine down a flexible hose from a funnel that the crewmember holds close to his or her body. Urine is stored temporarily and later dumped overboard. This process is simple, neat, and almost as fast as using the commode on the ground.

For solid wastes, we're held down on the toilet seat by padded, spring-loaded bars swung over the thighs. Once an airtight, sliding valve opens underneath the seat, a fan blows air and solid waste downward, away from the body, into the storage tank below.

After cleanup, the astronaut closes the sliding valve, which exposes the waste in the holding tank to the vacuum of space. This sterilizes and deodorizes the waste in the tank; it returns to Earth in a "freeze-dried" form.

On the space station, the current Russian-built toilet dumps urine overboard, but stows solid waste in airtight metal cans. The cans are stored onboard until they can be returned to Earth aboard a Progress cargo ship; the trash and waste burn up with the ship in the atmosphere.

The Least You Need to Know

- I flew on the space shuttle four times between 1994 and 2001.
- On my first two missions into space, we mapped the Earth's surface using a versatile radar laboratory.
- On my third mission, we researched the origins of the universe, and the potential commercial use of space, by launching two science satellites into orbit and returning them and their experimental results to Earth.
- My fourth mission was a crucial assembly mission to the International Space Station, called Alpha; my crew delivered the nerve center of the space station, the U.S.-built Destiny laboratory module.

NASA Today: Unlocking the Secrets of the Universe

In This Chapter

- ◆ A nonmilitary organization
- ◆ We put men on the Moon
- ◆ Teaching you
- ◆ Watching our budget

Before I talk about the NASA of today, the NASA that I worked for until my retirement in the summer of 2001, I want to give you a brief history of the organization. I'll go into NASA's history in much greater detail in Chapter 7, "Russia First Out of the Space Gate," but right now I'll just give you an outline.

Who We Are

The National Aeronautics and Space Administration (NASA) was born on October 1, 1958. It is a civil (that is, nonmilitary) organization designed to oversee the United States' exploration of space. It was created soon after the Russians put a satellite known as *Sputnik* into orbit around the Earth (see

Chapter 7). Fear that the Soviets' control of space would also give them a dangerously large military advantage, especially in strategic missiles and warheads, spawned an aggressive response in America and focused new attention on space exploration.

Dr. Jones's Corner

Although NASA scientists and engineers are responsible for many inventions that have changed day-to-day life, one that most people credit to NASA—Velcro—wasn't invented by the space agency at all. Velcro was actually invented by a Swiss inventor named George de Mestral in the early 1940s. After a walk with his dog, de Mestral noticed cockleburs stuck to his clothes and his dog's coat. Looking at the burs under a microscope, he saw stiff, hooked fibers all over their surface. De Mestral developed a fastening system with stiff hooks on one part, and soft loops on the other. NASA, although it had nothing to do with its creation, quickly discovered how useful Velcro was. In the near-weightless conditions in orbit, Velcro is perfect for temporarily holding objects and keeping them from floating away. "Velcro" is a combination of the words "velour" and "crochet." Velcro—don't leave Earth without it!

Mercury and Gemini

Right after NASA was created, the agency began planning on how to put people into space. NASA's first human spaceflight program was Project Mercury, aimed at finding out whether human beings, termed "astronauts," could survive and function while in *orbit* above the Earth. (See Chapter 11, "Catching Up: Yuri Gagarin, Alan Shepard, and Project Mercury," for more on Project Mercury.)

Space Talk

To **orbit** the Earth means to circle entirely around the globe in space. One time around the earth equals one orbit.

This pioneering effort was followed by Project Gemini (see Chapter 15, "Two Heads Are Better Than One: Project Gemini"), which sent a heavier spacecraft carrying two men into orbit around the Earth. With the successes of the Gemini program, America had whittled away the large lead that the Russians had taken during the early days of the space race.

Space Talk

The **seas** of the Moon, sometimes referred to as Maria, are not actually seas, and they never contained water. They are actually flat plains of ancient lava.

Apollo

Gemini was the forerunner of Apollo, President Kennedy's moonlanding program. Apollo was amazingly successful, putting men on the Moon—the first time in the *Sea* of Tranquility—and returning them to Earth a total of six times.

Following the Apollo program, the U.S. put a space station known as Skylab into Earth orbit, and with the Cold War thawing, even executed a joint orbital mission with the Soviet Union in 1975.

Space Shuttle

NASA's next challenge was the ambitious space shuttle program. The space shuttle is a reusable spacecraft that launches into space like a rocket, but returns to Earth on a runway, landing like a plane. For the last 20 years, it's been the most versatile spaceship on or off the Earth.

Robot Missions

In addition to the piloted space program, NASA has also been responsible for robot scientific probes that have taught us about the Moon, asteroids, and the planets.

Cosmic Facts

The Moon has approximately the same surface area as the continent of Africa.

The Hubble Space Telescope has made dramatic discoveries about the distant reaches of the universe, while the Viking and Mars Pathfinder craft have begun exploring the surface of Mars.

Revolutionary Satellites

Satellites launched by NASA have, through the years, revolutionized our abilities to communicate and predict the weather. Today, NASA remains committed to developing new technologies and research methods to continue mankind's ongoing investigation into our universe and to improve our lives back here on Earth.

Goals and Objectives

Today's NASA has goals and objectives that push the envelope of knowledge, searching out future needs and identifying technologies not yet imagined. Following are our four current goals.

Goal 1: Revolutionize Aviation

Enable a safe, environmentally friendly expansion of aviation.

Objective 1: Increase Safety. Reduce the aircraft accident rate by a factor of 5 within 10 years, and by a factor of 10 within 25 years.

Objective 2: Reduce Emissions. Reduce toxic emissions of future aircraft by 70 percent within 10 years, and by 80 percent within 25 years; reduce CO_2 (carbon dioxide) emissions by 25 percent and by 50 percent in the same timeframes.

Objective 3: Reduce Noise. Reduce the perceived noise of future aircraft by a factor of 2 within 10 years, and by a factor of four within 25 years.

Objective 4: Increase Capacity. Double the aviation system capacity within 10 years, and triple it within 25 years.

Objective 5: Increase Mobility. Reduce intercity door-to-door transportation time by half in 10 years, and by two thirds in 25 years. Reduce long-haul transcontinental travel time by half within 25 years.

Goal 2: Advance Space Transportation

Create a safe, affordable highway through the air and into space.

Objective 6: Mission Safety. Reduce the risk of crew loss by a factor of 40 within 10 years, and an additional factor of 100 within 25 years.

Objective 7: Mission Affordability. Reduce the cost of delivering payload to Earth orbit by a factor of 10 within 10 years. Reduce costs for both by an additional factor of 10 within 25 years.

Objective 8: Mission Reach. Reduce the time it takes to get to other planets by a factor of 2 within 15 years, and by a factor of 10 within 25 years.

Goal 3: Enable a Revolution in Aerospace Systems

Objective 9: Engineering Innovation. Develop the advanced engineering tools, processes, and culture to enable rapid, high-confidence, and cost-efficient design of revolutionary systems.

Objective 10: Technology Innovation. Develop the revolutionary technologies and technology solutions that enable fundamentally new aerospace system capabilities or new aerospace missions.

Goal 4: Commercialize Technology

Extend the benefit of NASA's research and technology.

Programs

Our research programs are based on our goals and objectives. All of our programs are planned and implemented in coordination with industry, Department of Defense, and Federal Aviation Administration partners to ensure that program deliverables are transferred in a timely manner for further development and implementation.

Cosmic Facts

If there was a highway from Earth to the Sun, it would take, staying within speed limits, 170 years to make the trip. Be sure to bring the sunscreen!

Here are our current space and aviation research programs:

- Space Launch Initiative
- Advanced Space Transportation Program
- Small Aircraft Transportation System
- Aviation Operations Systems
- Intelligent Synthesis Environment
- Vehicle Systems Technology
- Information Technology
- Quiet Aircraft Technology
- Propulsion and Power
- Intelligent Systems
- Flight Research
- Rotorcraft

The technologies that NASA hopes to develop are long-term, high-risk, and high-payoff. It's our job to take risks, to push technology, to gamble for a big return, in keeping with a prudent assessment of what we may lose. Progress doesn't come easily.

Education

NASA maintains a close working relationship with the educational community. At the high-school level, NASA works closely with educators to provide supplementary educational programs, products, and services that are valuable resources for teachers and students; our efforts support the national standards for education. NASA also depends on colleges and

Cosmic Facts

NASA's Aerospace Technology education outreach programs offer students a wide range of opportunities to learn about math, science, and technology. The agency ties these subjects into our exploration and aviation programs, motivating students to master skills needed for a future technical career.

universities for research support and intellectual expertise, while the university research community benefits from NASA resources and direction.

Distance Learning Expeditions

NASA offers students an interactive video event known as a Distance Learning Expedition. Using a live teleconference, astronauts come into the classroom to teach youngsters about a variety of space subjects. I've led a number of these Expeditions, taught from Johnson Space Center. Each product has been designed using sound instructional principles and satisfies national education standards. Here are some of the topics offered, and the grade levels they are designed for:

- Space Farming, Grades 5–12
- Rocks from Space, Grades 4–8
- X-38 (Astronaut Lifeboat), Grades 3–12
- Space Food, Grades 3–8
- Robotics, Grades 5–12
- BIO-Plex, Grades 5–8
- Effects of Space on the Human Body, Grades 4–12
- Microgravity, Grades 5–8
- Globes, Maps, and International Space Station, Grades 3–4
- Imagery from Space, Grades 5–8
- Careers, Grades 7–12
- Astronomy, Grades 5–12
- International Space Station Virtual Tour, All Grades

Teachers plan in advance to participate in each seminar, and students engage in preparatory learning before undertaking each expedition. (Please contact the Johnson Space Center for more information—the experts there have all the technical requirements of the Expedition video teleconference equipment.)

Realities: The Need to Stay on Budget

NASA's funding has been tight for more than a decade. NASA's budget for Fiscal Year 2002 is about $14.6 billion, which is only about 0.7 percent of Federal expenditures. By contrast, during the Apollo program of the 1960s, the nation spent more than four percent of its budget on civilian space exploration.

Cosmic Facts

When NASA was first formed, it was in charge of every civilian rocket launch in the United States, but that is no longer the case. Starting in 1989, NASA began to contract out for launch services, thus boosting the still-growing U.S. commercial launch industry. Many launch operations are now in the hands of the manufacturers of the vehicles and the U.S. Air Force. (The only U.S. manned space vehicle is the space shuttle, which remains solely under NASA's auspices.) Some of these non-NASA launch vehicles are the Delta II (the Boeing Company); Atlas I and Atlas II (General Dynamics Space Systems Division); and the Titan IV (Lockheed Martin Space Launch Systems). NASA retains oversight responsibilities for those vehicles that carry NASA payloads, but more and more of them will be launched by commercial space launch companies.

NASA has responded by becoming more efficient: reducing its workforce, cutting the cost of operating the space shuttle nearly in half, slowing space station construction, and asking our international partners to contribute more to station construction costs. As of this writing, NASA has its hands full with space station construction, solar system exploration, and continued flights of the shuttle. If we're going to return to the Moon or set off for the asteroids or Mars, we'll need public support for more funding.

The Least You Need to Know

- NASA operates all American civilian space programs.
- The space agency is chartered to explore the atmosphere and space, and to share its results with the public and the scientific community.
- NASA was born at the height of the Cold War but continues today to push the boundaries of exploration and knowledge.
- Less than a penny of every federal dollar spent goes to space exploration.

What Makes a Rocket Go Up?

In This Chapter

- ◆ Newton's Laws of Motion
- ◆ Built in stages
- ◆ Why things go into orbit
- ◆ The founding fathers of rocketry

Welcome to Rocketry 101. In this chapter, I'll take you through the fundamentals of rocket science. We'll zip through a brief history of rockets, get a feel for how they work, and learn about the men who made rocketry what it is today. Our story starts some 800 years ago …

From China to the Fourth of July: A Brief History of Rocketry

Rockets probably got their start as fireworks, used in celebrations, but rockets quickly evolved into weapons. Known as "fire arrows," they were used in China by the Mongols in the thirteenth century. Rockets as weapons are even an instrumental part of our American history. Remember the "rockets' red glare" from our National Anthem? Those were British "Congreve" rockets, launched at Ft. McHenry in Baltimore—my hometown. And what do we do

Space Talk _____

In 1379, an Italian named Muratori used the word *rochetta* when he described types of gunpowder-propelled fire arrows used in medieval times. This is believed to be the first use of the word later translated in English as **rocket**.

Cosmic Facts _____

The work "Chronicles" by Jean Froissart, published in 1410, outlined the design of tube-launched military rockets.

on our Independence Day, also known as the Fourth of July? We set off rockets, that's what. Rockets went on to be used as life-saving and signal devices, but weren't really effective as weapons until World War II.

First Rocketeer

The first attempt to propel a man using a rocket may have come in the sixteenth century when, according to a Chinese folktale, a man named Wan-Hoo attempted a blast-off. He reportedly took two large horizontal stakes and tied a seat between them. Under the primitive device were placed 47 rockets, to be lit all at the same time.

When the rockets were ignited, they burned erratically and could not provide enough effective thrust to move the contraption. Wan-Hoo reportedly burned to death in the resulting fire.

Newton's Theories

The theories of Sir Isaac Newton, published during the seventeenth century, were instrumental in the development of rockets. Here are some examples:

♦ Newton's Third Law of Motion stated, "For every action there is an equal and opposite reaction." This remains the fundamental principle of rocket propulsion, even today known as "reaction thrust." As the flame and hot gases rush downward out of the opening at the base of the rocket, the resulting reaction pushes the rocket upward.

♦ Newton theorized that if an object could be fired fast enough and at a high enough altitude, it would not fall back to the ground, but could achieve an orbit around Earth.

♦ Newton theorized that a force called "gravity," proportional to the mass of a body and diminishing with its distance from another body, acted on the Sun and the planets and controlled their orbits. The resulting system of orbital mechanics successfully predicted small, natural changes, or perturbations, in the orbits of the known planets.

♦ The Newtonian "system of the world," which predicted previously unexplained variations in the orbital paths of the planets, led directly to the discovery of Neptune in 1846 and Pluto in 1930.

◆ Jet propulsion, using Newton's Third Law of Motion, was demonstrated successfully in 1720 when Dutchman Jacob Willem Gravesande built a model of a car propelled by an internal steam engine.

The rocket's potential as a weapon reached maturity during and following World War II. All the major powers used small rockets as a form of artillery, and fighter-bombers used rockets as an effective weapon for ground attack. Nazi Germany marshaled a huge effort to develop the guided missile as long-range artillery; the V-2 rocket (see Chapter 6, "Wernher von Braun and Early U.S. Rockets") became the forerunner of the intercontinental ballistic missiles of today. Modern precision-guided rockets were used in the Gulf War both from the air and as ground-based artillery. In 2002, the civilian descendants of Cold War missiles are still the mainstay of the commercial space launch industry.

How a Rocket Works

Rockets produce the hot gases that thrust them forward by burning two ingredients in their combustion chambers—a fuel and an oxidizer. Jet engines combine their fuel with oxygen in the air to produce combustion. For a rocket to function in the upper atmosphere and in the vacuum of space, rockets must carry not only fuel, but an oxidizer as well: usually oxygen, or an oxygen-containing chemical.

There are two basic types of rockets, distinguished by the form of the fuel and oxidizer used. Early rockets and today's fireworks use a form of gunpowder, a solid material, as fuel. These solid-fuel rockets have the advantage of being storable; their disadvantage is that once started, they cannot be shut down. The space shuttle's *booster rockets* are an example of a solid-fuel rocket. They burn a solid propellant mix of aluminum oxide as a fuel and ammonium perchlorate as an oxidizer.

Liquid-fuel rockets carry their fuel and oxidizer in liquid form. They can be shut down and restarted by controlling the flow of propellants through valves and pumps. Also, liquid propellants pack more energy than their solid-fuel counterparts, and so liquid-fuel rockets are more efficient in hauling weight into orbit. The World War II V-2 rocket was a liquid-fuel rocket, burning alcohol as a fuel and liquid oxygen as its oxidizer. The space shuttle main engines burn liquid hydrogen as a fuel, combined with liquid oxygen as the oxidizer.

Space Talk

Booster rockets are designed specifically to boost a spacecraft into Earth orbit or beyond.

Solid-fuel rockets are usually ignited by a small powder charge that sends a flame shooting down a hollow tube at the rocket's core. The heat ignites the solid propellant and sends exhaust gases racing out the nozzle.

Cosmic Facts

The French are reported to have made extensive use of war rockets throughout the fifteenth century. In 1429, French troops led by Joan of Arc reportedly used rockets in their successful defense of the city of Orleans.

Dr. Jones's Corner

Nathaneal Lye published "The Art Of Gunnery" in 1647, which contained detailed instructions on the construction, handling, and operation of a variety of pyrotechnic and military rockets.

To ignite a liquid-fuel rocket engine, fuel from one tank and oxidizer from another enter the rocket's combustion chamber and mix. The mixture starts burning when ignited by a small charge of gunpowder, an electric spark, or a chemical reaction between the propellants.

The liquid oxygen, or other oxidizer, is necessary to produce a steady output of hot exhaust gases. There is not enough oxygen in the air to keep burning the fuel at such a rapid pace—and the need for a self-contained oxygen supply only increases as the rocket gets higher, because the air gets less and less dense with altitude.

Just as Newton would have predicted, the hot gases released by the fire in the combustion chamber push outward and flow out the nozzle at the rocket's end. This pushes the rocket in the opposite direction and—if everything's working right—upward.

The lifting power of the rocket varies depending on the quantity and type of fuel. That lifting force is called the rocket's "thrust," usually measured in pounds or kilograms. For example, the space shuttle's boosters and main engines put out around seven million pounds of thrust at lift-off. The Saturn V booster that sent humans to the moon produced 7.5 million pounds of thrust at blast-off.

Built in Stages

Rockets work because the thrust they produce is greater than the weight of the rocket—the force exerted on it by gravity. It follows that the lighter the rocketship, the less energy it takes to push it upward. That's why most modern rockets are built in what are called stages. In other words, there are a series of rockets all stacked on top of one another.

Space Talk

Payload is a launch vehicle's cargo—that which is to be put into Earth orbit or propelled into deep space.

The bottom rocket—the first stage—ignites first, of course. When the first stage is out of fuel, it is jettisoned—that is, cut loose—from the rest of the spacecraft. The rocket immediately above that one then ignites, and with the weight of the lower stage gone, the spacecraft is lighter and more efficient than ever.

As you can see, a rocket is never heavier than when it is on the launchpad. The second stage does not have to drag upward the weight of the first stage's engines and empty fuel tanks. When the second stage runs out of fuel in turn, it is jettisoned and the third stage takes over. This process continues until only the *payload* is left, with enough speed to orbit the Earth or leave for the planets.

Why Objects Go into Orbit

Think of a spacecraft as a weight on the end of a string. When you whirl that weight around your head, the weight constantly wants to go in a straight line (Newton's First Law) but cannot because the string holds it back and forces it to go around in circles.

The speed that you imparted to the weight with your wrist and arm is just like the speed given to a spacecraft by its rocket engines. The string in this analogy is acting like the force of Earth's gravity, which will not allow the spacecraft to go in a straight line, but instead keeps the object going around and around the Earth—in orbit. On the space shuttle I travel at five miles per second, and without the pull of Earth's gravity, my crew and I would race off across the solar system at that speed.

Orbital velocity is about 17,500 miles per hour for typical shuttle and station orbits. Yes, launch direction does matter. Each point on the Earth rotates eastward, with a speed depending on latitude. The eastward speed at the North Pole is zero; at the equator, over 1,000 miles per hour. So we do get a boost from Earth's rotation, and it's a maximum if we launch straight east at the equator. That's why we launch east from Cape Canaveral if possible. A polar-orbit launch, one that sends a satellite on a path that takes it over both poles of the Earth, gets no benefit from the spin of the Earth, so we get less payload to orbit on a polar launch with a given size rocket. There's also a penalty in payload if we launch to the Space Station's 51.6 degree orbit (it's not equatorial because we have the Russians as partners, and due to their latitude and China's border, they launch into an orbit inclined 51.6 degrees to the equator).

There are also medium Earth orbits (MEO), above 5,000 nautical miles (5,758 statute miles), where radiation exposure is acceptable for astronauts, and high Earth orbits that go out a substantial portion of the way to the Moon, say 100,000 miles (that's where the satellite *Chandra* went). A GEO, or geosynchronous orbit, is about 22,300 statute miles up, at a point where the orbit's period is the same as Earth's rotation period—24 hours.

Space Talk _____

The perpetual path of a planet around a Sun, or the path of a moon around a planet, is called its **orbit**. When a spacecraft circles a planet or moon it is said to be in orbit. The speed of a spacecraft determines its orbital height, and its initial direction on launch determines the shape of its orbit. Once a spacecraft is in orbit, no further rocket engine firings are needed, except to change the orbit's shape, or counteract other forces, like drag from the atmosphere.

Size of Payload

The larger the payload, of course, the larger the rocket needed to place it in orbit. In fact, the payload weight, destination, and purpose determine what vehicle capabilities are required for each mission.

Space Talk

Near-Earth orbit is also called "low Earth orbit." Atmospheric drag won't permit flights below 100 nautical miles (115 statute miles), and because of the Van Allen radiation belts, we don't send people or satellites into orbits above 300 nautical miles (345 statute miles). Of course, satellites can function above the belts and beyond.

A low-weight spacecraft designed to operate *in near-Earth orbit* might be flown aboard the comparatively small Pegasus or Taurus.

Sending a manned Apollo spacecraft to the Moon required the massive Saturn V. The powerful Titan-Centaur combination carried large and complex unmanned scientific explorers such as the Vikings and Voyagers to examine other planets. Atlas/Agena sent several spacecraft to photograph and then impact or land on the Moon. Atlas/Centaurs and Deltas have launched more than 260 spacecraft for a wide variety of purposes, covering the broad range of the national space program.

The following chart shows the amount of thrust at lift-off for various United States rockets.

Rocket	Thrust
Scout	132,240 lbs.
Pegasus	140,000 lbs.
Pegasus XL	200,000 lbs.
Atlas/Agena	468,500 lbs.
Taurus	495,400 lbs.
Delta	873,400 lbs.
Saturn 1B	1.6 million lbs.
Titan-Centaur	2.4 million lbs.
Saturn V	7.5 million lbs.

How Do Satellites Stay in Orbit?

Two things are needed to get an object into and keep it in orbit. First, the object must reach the appropriate height, at the very edge of the Earth's atmosphere. Also, it must have the velocity to maintain itself in orbit. For example, a balloon, assuming one could float one that high up, would not achieve orbit no matter how high it got, because it would not be moving fast enough.

If something slows down an orbiting satellite, it will fall back to Earth eventually. The speed is needed to carry it around the curving surface of the Earth. Going too slowly will result in an eventual curve down into the atmosphere, where it will burn up.

But if the speed is just fast enough, the satellite will fall—orbit—in just the same shape curve as the surface of our Earth. Eventually the small amount of *drag* from the thin upper atmosphere will tug a satellite down to Earth. But, because atmospheric friction is barely noticeable in space, this "orbital decay" takes a while, so satellites can stay in orbit for many years.

Cosmic Facts

Objects burn up upon entering Earth's atmosphere because of friction. Because of the speed of the falling object, this friction caused by the object moving through, thus rubbing against, the atmosphere, heats the object up until it burns.

Space Talk

Even though the atmosphere surrounding a spacecraft in Earth orbit is very thin, it does exist. There is enough atmosphere to cause a small amount of friction as the spacecraft moves through it. This force is called **drag,** and it can decrease a spacecraft's speed.

The Founding Fathers of Space Travel

It took many brilliant thinkers, scientists, and engineers to transform rocketry from a tactic of Chinese warfare to men walking on the Moon. Here are a few brief bios of those who led the way.

Hermann Oberth: Romanian Visionary

The top rocket scientists and engineers of the mid-twentieth century counted as their major influence one Hermann Oberth, a Romanian math teacher living in Germany who became fascinated with rockets following World War I. Oberth put forth the theory that liquid fuels, such as gasoline and liquid oxygen, would cause rockets to go higher and farther than rockets using solid fuels, such as gunpowder. Oberth theorized that a 400-metric-ton rocket, using liquid fuels and carrying two men, could be successfully launched into Earth orbit.

In 1923, Oberth wrote a slender but influential book titled *The Rocket into Interplanetary Space*. When in 1929 the legendary German filmmaker Fritz Lang made a science fiction movie called *The Girl in the Moon*, he hired Oberth to be his technical advisor. Although Oberth was quite helpful with the special effects showing a rocket launching in the film, their attempt to launch a real rocket as a publicity stunt to promote the movie was a dismal failure. Rocket capabilities still lagged well behind science fiction.

Konstantin Tsiolkovsky: Russian Prophet

The first Russian scientist to make innovations in the field of rocketry was Konstantin Tsiolkovsky. Born in the mid-nineteenth century, he studied astronomy as a child and later, working as an inventor, tried to build an airship. Even before Hermann Oberth, Tsiolkovsky suggested that liquid rocket fuels were preferable to solids and suggested that liquid oxygen and liquid hydrogen were the fuels of choice. Tsiolkovsky is revered by the Russians as the father of their space effort, in their mind eclipsing Robert Goddard.

Robert Goddard: Inventor

Robert Goddard, from Worcester, Massachusetts, was a reclusive doctor of physics who became interested in outer space after reading H. G. Wells's *War of the Worlds*. Goddard applied for and received a grant from the Smithsonian Institution to support his research into flight and produced a ground-breaking 69-page book called *A Method of Reaching Extreme Altitudes*.

In March 1926, Goddard launched the first liquid-fueled rocket, which flew 184 feet. He was the first to use a gyro as a primitive guidance system, allowing the rocket to detect when it was deviating from the vertical, making corrections using heat-resistant vanes that deflected the rocket's exhaust.

In 1935, Goddard built a rocket that reached an altitude of 7,500 feet. This was the peak of his efforts. Five years later he experimented with even larger rockets, but these all became unstable right after lift-off and crashed nearby.

Goddard worked in rocketry for the military during World War II, but his innovations were relatively minor. He died of cancer in August 1945.

Sergei Korolev: The "Chief Designer" of the Russian Space Program

The head of the Soviet space effort, the first to put a man in orbit, was Sergei Korolev. He was a talented aeronautical engineer, who, ironically, had been for years a political prisoner of Soviet dictator Joseph Stalin. It was Korolev who built the first successful liquid-fueled Russian rocket, and who remained, for many years, the "chief designer" of Soviet missiles. The Russians had previously used millions of solid-fueled battlefield rockets in World War II.

That first rocket was called the "09." It reached 1,300 feet before crashing earthward in August 1933. Five years later Korolev was swept up in Stalin's purges (in which just about everyone who had been critical of Stalin or his regime was either imprisoned or executed). Korolev continued working on his rocket theories even in prison. He was released in 1944 to aid in the desperate fight against the Nazis. Following World War II, Korolev was sent to Berlin to learn as much as possible about the V-2 rocket.

As with the Germans, the object of Korolev's work was not the exploration of space, but rather development of an intercontinental ballistic missile, or ICBM—a missile that could carry a nuclear warhead from one continent to another, causing mass destruction.

At the end of the Second World War, Stalin had a powerful army, but lacked sea and air power. He especially feared the U.S. strategic bomber force, equipped with atomic weapons. Stalin believed that having a ballistic missile system could be the great equalizer.

Cosmic Facts

Throughout what was to become the space race, the U.S. scientists had a great advantage over their Russian counterparts in that their budget was always much larger. The Russians often had to work under spartan conditions, without adequate budget to test components properly.

James Wyld and the American Rocket Society

The American Rocket Society started in 1930 as a group of science-fiction aficionados, but soon developed into an organization that sponsored real research into rocket science. Their first experiments were less than successful, but they grew steadily more proficient.

One of the society's most important members was a Princeton student named James Wyld, who was experimenting with ways to cool rocket engines more efficiently. After graduation, Wyld moved to New York and eventually developed the engine for the rocket-powered Bell X-1, the first aircraft to break the sound barrier. The ARS endorsed the notion of a nonmilitary organization to oversee the United States space program, thus giving a boost to the eventual birth of NASA. The society switched its research emphasis to engineering with the coming of World War II, and in 1963 it merged with the Institute of the Aeronautical Sciences to form the American Institute of Aeronautics and Astronautics.

Theodore von Karman and the Jet Propulsion Laboratory

Theodore von Karman was a professor of aeronautics at the California Institute of Technology, more commonly referred to as Caltech, during the 1930s and '40s. His work in aerodynamics, jet engines, gas turbines, and wings for high-speed flight led to the creation of the Jet Propulsion Laboratory, where state-of-the-art research and development in space technology has been going on ever since.

And then there was Wernher von Braun, the man who designed the German V-2 missile, and who later became the most influential designer of U.S. missiles and space boosters following World War II. We'll discuss him in Chapter 6.

The Least You Need to Know

- For most of their history, rockets were used as weapons—and were seldom effective.
- Newton's Laws of Motion were key to the future development of rocket science.
- Pioneers from several nations helped give birth to the Space Age.
- The larger the payload, the more fuel—and the bigger the rocket—needed to put it in orbit.

Chapter 4

So You Want to Be an Astronaut?

In This Chapter

- ◆ Things you'll need
- ◆ Applying
- ◆ Message to the young
- ◆ Other great NASA careers

The first astronauts were all military test pilots, who'd met stringent height and experience requirements. Over the last 35 years, the requirements have changed considerably.

Requirements to Be an Astronaut

With the advent of the space shuttle 20 years ago, scientific and engineering applicants no longer are required to be pilots. Today, NASA is looking for candidates with broad technical experience.

I like to tell people that we have two "flavors" of professional astronaut today. The first category is "pilot astronaut." These are military-trained test pilots who are responsible for getting the shuttle to and from orbit and for maneuvering the vehicle in space. The second "flavor" is a *mission specialist* astronaut, who isn't necessarily military trained. There is also a class of noncareer astronauts called *payload specialists*.

> **Space Talk**
>
> **Mission specialists** are scientists and engineers who operate the shuttle's science payloads, handle the robot arm, and don spacesuits to work outside the shuttle and station on spacewalks.

Of course, we also fly international partner astronauts on NASA's space shuttle and on the NASA-led International Space Station. Each partner country has its own astronaut selection criteria, certified by NASA trainers.

Minimum Requirements

There are a few minimum requirements you have to meet before you can even be considered to be an astronaut. Here are the basic qualification requirements to enter NASA's Astronaut Candidate Program. To be a Mission Specialist Astronaut Candidate, you must have:

> **Space Talk**
>
> A **payload specialist** is an expert chosen by NASA to fly—on one or two flights—when a flight experiment or activity demands his or her specialized training. Payload specialists receive basic spaceflight training and train with their shuttle crews for about a year before flight.

- ♦ A bachelor's degree from an accredited institution in engineering, biological science, physical science, or mathematics. Your degree must be followed by at least three years of related, progressively responsible, professional experience. An advanced degree is desirable and may be substituted for part or all of the experience requirement (master's degree equals one year, a doctoral degree equals three years).

- ♦ The ability to pass a NASA class II space physical, which is similar to a civilian or military class II flight physical and includes the following specific standards: distant visual acuity—20/150 or better uncorrected, correctable to 20/20, each eye; blood pressure: 140/90 measured in a sitting position.

- ♦ A height between 58.5 and 76 inches.

To become a Pilot Astronaut Candidate, you must have:

- ♦ A bachelor's degree from an accredited institution in engineering, biological science, physical science, or mathematics. The degree must be followed by at least three years of related, progressively responsible, professional experience. An advanced degree is desirable. Quality of academic preparation is important.

♦ At least 1,000 hours of pilot-in-command time in jet aircraft. Flight test experience highly desirable.

♦ The ability to pass a NASA Class I space physical, which is similar to a military or civilian Class I flight physical and includes the following specific standards—Distant visual acuity: 20/50 or better uncorrected, correctable to 20/20, each eye; blood pressure: 140/90 measured in sitting position.

♦ A height between 64 and 76 inches.

Cosmic Facts

Applicants for the Astronaut Candidate Program must be citizens of the United States.

Education Requirements

Applicants for the Astronaut Candidate Program must meet the basic education requirements for NASA engineering and scientific positions—specifically: successful completion of a standard professional curriculum in an accredited college or university, leading to at least a bachelor's degree, with major study in an appropriate field of engineering, biological science, physical science, or mathematics.

The following degree fields, while related to engineering and the sciences, are *not* considered as qualifying:

♦ Degrees in technology (Engineering Technology, Aviation Technology, Medical Technology, etc.)

♦ Degrees in Psychology (except for Clinical Psychology, Physiological Psychology, or Experimental Psychology, which are qualifying)

♦ Degrees in Nursing

♦ Degrees in social sciences (Geography, Anthropology, Archaeology, etc.)

♦ Degrees in Aviation, Aviation Management, or similar fields

Dr. Jones's Corner

We've begun to see the dawn of the tourism era in space. Over the next few years, more and more ordinary civilians (with necessarily deep pockets) will fly on the Russian *Soyuz* spacecraft, or perhaps even aboard the shuttle.

How to Apply

When NASA decides to select additional astronaut candidates, consideration is given only to those applications on hand on the date the decision is made. Applications received after that date are retained and considered for the next selection.

Applicants are notified annually of the opportunity to update their applications, and to indicate continued interest in being considered for the program.

Cosmic Facts

For civilians, the application package may be obtained by writing to: NASA Johnson Space Center/Astronaut Selection Office/ATTN: AHX/Houston, TX 77058. Civilian applications will be accepted on a continuing basis.

Cosmic Facts

Applicants should be aware that selection as an astronaut candidate does not ensure selection as an astronaut. Final selection depends on satisfactory completion of the one-year training and evaluation period. Civilian candidates who successfully complete the training and evaluation and are selected as astronauts become permanent Federal employees and are expected to remain with NASA for a period of at least five years.

Those applicants who do not update their applications annually are dropped from consideration, and their applications aren't retained.

After the preliminary screening of applications, additional information may be requested for some applicants, and persons listed on the application as supervisors and references may be contacted.

Military personnel on active duty must submit applications to their respective military service and not directly to NASA. Application procedures are disseminated by each service.

Personal interviews and thorough medical evaluations are required for both civilian and military applicants under final consideration.

Once final selections have been made, all applicants who were considered are notified of the outcome of the process. Selection rosters established through this process may be used for the selection of additional candidates during a one-year period following their establishment.

Selected applicants are designated "Astronaut Candidates" and are assigned to the Astronaut Office at the Johnson Space Center, Houston, Texas. The astronaut candidates undergo a one-year training and evaluation period, during which time they are assigned technical or scientific responsibilities allowing them to contribute substantially to ongoing programs. They also participate in the basic astronaut training program, which is designed to develop the knowledge and skills required for formal mission training upon selection for a flight. Pilot astronaut candidates must maintain proficiency in NASA aircraft during their candidate period.

Civilian candidates who are not selected as astronauts may be placed in other positions within NASA depending upon Agency requirements and manpower constraints at that time. Successful military candidates are detailed to NASA for a specified tour of duty.

For a Well-Grounded Career

Not all the people who work at NASA are astronauts, of course. For those interested in working at NASA who would prefer to keep their feet on the ground, here is a list of potential occupations:

Scientists	Engineers	Mathematicians
Astronomer	Aeronautical	Computer Scientist
Biologist	Aerospace/Astronautics	Mathematician
Chemist	Biomedical	Statistician
Geologist	Chemical	Systems Analyst
Medical Doctor	Civil	
Meteorologist	Computer	
Nutritionist	Electrical	
Oceanographer	Environmental	
Physicist	Industrial	
Physiologist	Materials	
Psychologist	Mechanical	
Veterinarian	Nuclear	
	Petroleum	
	Plastics	
	Safety	
	Systems	

Tips for Youngsters

I've wanted to be an astronaut since I was about 10 years old. There isn't a magic formula for school courses that will prepare you for astronaut training. But it's a safe bet that nearly any type of science, math, or engineering course is good training.

Science

You should pursue any science courses that interest and engage you, because almost all of them have a connection to space exploration. We have astronauts who are specialists in fields ranging from astronomy to zoology, and NASA needs a broad array of experts. Choose something you love, and then work hard to excel.

Math

Math is the key to success in the sciences and engineering. It's the common language, if you will, of the sciences. By mastering your math courses, you'll be prepared to pursue any technical field with confidence. Take a math course every year in high school and college. You can't get too much of it.

Team Sports

Space shuttle and space station crews have to be expert at getting along with others and working as a team. Team sports teach not only how to be a strong competitor, a necessary trait for success in life, but also how to get the most out of a group of individuals in a team effort. Teamwork, and reliance on each other, will be critical for those first crews heading off to Mars or back to the Moon.

Cosmic Facts

I changed majors three times in college—from aerospace engineering, to physics, and finally to basic sciences. I took a math course every semester, along with courses in astronomy and space sciences.

Tips for College Students

College students can choose almost any major (see the preceding table) that ties in to space exploration. Choose a major field that you enjoy, and then run with it to the best of your ability.

Work at excellence—aim to graduate with high marks, and look for a career opportunity that will get you experience working with others on a technically demanding project. NASA is looking for people who can work together to solve real, complex problems.

Tips for Those Entering the Military

I left the military 18 years ago, but I've worked with active duty military colleagues all through my astronaut career.

NASA values the operational, real-world experience brought to the agency by its military members. Those entering the military should plan to advance their formal technical education while in uniform, aiming for at least a master's degree. I advise choosing a career specialty in engineering, science, or space operations.

Space Cowboys, Cowgirls, and the Final Frontier

In my 11 years as an astronaut, I worked with a dazzling array of bright people. Not all of them had a lifelong dream of becoming a space traveler, but all are now caught up in the excitement of space exploration. They want to be in the forefront of our adventure—our destiny—in space. All of them, from test pilots to space scientists, are dedicated to bringing the benefits of the future down to Earth to us today.

So if you are a young person considering a career with NASA, but you are afraid that it isn't all that it's cracked up to be, you can stop worrying now. There are few things you can do in life as exciting and rewarding as being on the "cutting edge" of space exploration.

> **Dr. Jones's Corner**
>
> NASA has an affirmative action program goal of having minorities and women among those qualified as astronaut candidates. Therefore, qualified minorities and women are encouraged to apply.

The Least You Need to Know

- The three types of professional American space travelers are pilot astronauts, mission specialists, and payload specialists.
- Astronauts can apply from either the military or the civilian worlds.
- Astronaut candidates undergo a one-year training program, leading to selection as an astronaut and an eventual flight assignment.
- Excellence in a scientific or engineering education, beginning in high school, is the key to becoming a successful astronaut candidate.

Visiting NASA: Y'all Come On Down, Hear?

In This Chapter

♦ Kennedy Space Center
♦ Johnson Space Center
♦ Ames Research Center
♦ Other NASA centers to visit

All NASA's installations were built with American taxpayers' dollars. Therefore, every U.S. citizen owes it to him- or herself to come down and view the good things those tax dollars have bought. Seeing the people and hardware that are carrying out the dream of centuries is both exciting and uplifting. It's my hope that readers of this book are curious enough about NASA to visit us. This chapter is dedicated to making that visit as pleasant as possible.

Kennedy Space Center

Kennedy Space Center (KSC) is America's spaceport. Here the space shuttle blasts off and returns to a gentle landing; here the International Space

Cosmic Facts

If you wish to visit KSC *on business*, you must get a clearance badge that grants access to the space center or Cape Canaveral Air Station. Clearance to enter the space center complex is usually arranged by the person or office you're visiting.

Cosmic Facts

Approximately 6,000 acres of land have been cleared for KSC operations. However, KSC facilities and personnel are spread over a much larger area. The entire complex is part of the Merritt Island National Wildlife Refuge, and visitors can routinely see alligators, bald eagles, egrets, and manatees. In fact, NASA has had to contend with launch pad woodpeckers that have chipped holes in the insulation of the shuttle's external fuel tank.

Station's components are checked out for launch. In the past 30 years, KSC has played host to millions of tourists from all around the U.S. and the world. Visit the KSC, and hear and see the story of how the U.S. built a space program that launched men to the Moon, orbited satellites that have improved our lives, and sent probes into distant space to unravel the mysteries of the universe.

What to See

KSC is located near Cape Canaveral on the east coast of Florida, approximately midway between Jacksonville and Miami. KSC is built on 140,000 acres of land, swamp, and waterways. The Center includes one of the world's longest runways and the largest building in the U.S. If you're interested in the beginning or end of a space shuttle mission, or the history of our space program, then a tour of KSC will seem like a dream come true.

From the KSC Visitor Complex, a bus tour is available that takes you back to NASA's beginnings. On that tour you'll see the launch structures that were used to launch the Mercury, Gemini, and early Apollo missions as well as the pads that sent the Mariner, Explorer, Viking, and Voyager spacecraft to the planets. You can also visit the Air Force Space and Missile Museum at Cape Canaveral Air Force Station.

Dr. Jones's Corner

If you want to guarantee yourself a look at one of the most impressive sights known to man—the launching of the space shuttle—try to schedule your arrival just before a launch, so you have the time to hang around for a few days in case of weather or technical delays. I always tell my friends who are coming to a launch to plan for a fun Florida vacation—and if they're lucky, they'll get to see a shuttle launch, too!

The KSC Visitor Complex is open every day of the year (except Christmas Day and certain launch days). The last KSC tour departs about four hours before closing. The Complex is on State Road 405, on the east bank of the Indian River. If you're coming from Orlando, take Route 528, often called the Bee Line Expressway, to U.S. 1 North, to State Road 405.

When visiting KSC, you can go on a journey that takes you from the Vikings' discoveries of Greenland and Iceland to the landing of the Viking Lander, the first U.S. probe to land on Mars, in 1976. You'll have the opportunity to see and actually touch a piece of Mars, which fell to the Earth as a meteorite. Guests may also submit their names to fly in space on a future mission! Walk through full-scale mock-ups of space station modules and visit the viewing gallery, where you can see actual space station components being readied for flight. Catch a flick at the IMAX theater. See a three-stage Saturn V moon rocket, along with a dramatic recreation of the first manned Saturn V launch. With hands-on exhibits and actual Apollo hardware, the Apollo/Saturn V Center brings to life the U.S. space program's missions to the Moon. While riding an air-conditioned bus, you'll see the Apollo/Saturn V Center, the LC 39 Observation Gantry, and the International Space Station Processing Facility.

See the rockets that launched chimps, astronauts, and robots into space, including a Mercury-Redstone, similar to the one that carried Alan Shepard into space. Another special vehicle in the collection is an awe-inspiring Mercury-Atlas booster, identical to the rocket that carried John Glenn into space as America's first man in orbit.

Among the rockets you'll see displayed at the Kennedy Space Center Visitor's Complex are a Mercury-Atlas, like those ridden by John Glenn and Scott Carpenter, and a Mercury-Redstone, the booster that propelled Alan Shepard and Gus Grissom on their sub-orbital hops in 1961.

(Photo by MacIntyre Symms)

See a Launch

If you're going to be in Florida for a scheduled launch of the space shuttle, be sure to get a launch guest pass from NASA. The passes are free, and they enable you to watch the launch from KSC's NASA Causeway, which is about five miles from the launch pads. Passes are limited, so check for them at www.ksc.nasa.gov under the area marked "How to view a shuttle launch."

The KSC Visitor Complex also sells launch-viewing passes, which include a bus ride to a viewing area. Visitors can choose either option, but the KSC Causeway passes discussed above are free as long as they are available.

But don't fret if those tickets are sold out, as there are many great places in the area outside KSC proper to watch the launch. Try Jetty Park in Cocoa Beach, or the shoreline fronting Titusville.

> ### Dr. Jones's Corner
>
> You should plan on spending at least one full day at the Visitor's Complex. Bus tours can fill an entire morning and afternoon, and then there are films, exhibits, and artifacts. And, if you time your vacation properly, you can watch a launch from the KSC Causeway.

> ### Cosmic Facts
>
> If you get the opportunity to see a launch at KSC, there are a couple of items you should remember to bring: sunscreen and insect repellent. If there are two things the Cape has plenty of, they are sunshine and mosquitoes.

Space Camp

U.S. Space Camp is a five-day program jam-packed with astronaut training experiences for young people. Activities include simulated space shuttle missions, IMAX movies, training simulators (like the 1/6th Gravity Chair), model rocket building and launches, hands-on scientific experiments, and informal talks on the past, present, and future of space exploration. Kids must already be attending classes in the fourth grade and be at least nine years old. The only other requirement is the desire to have a great time!

Kids stay in bunk-bed style dorm rooms in any of three locations:

- Alabama (Huntsville)
- California (Mountain View)
- Florida (Titusville)

Also available are programs for older students: Space Academy for 12–14-year-olds, and Advanced Space Academy for 15–18-year-olds. Another popular program is Parent/Child Space Camp, a weekend of activities and missions where parents and kids go through the same program together.

Space Camp has been operating since 1982. It is the largest U.S. camp operation, having graduated almost 300,000 campers. Space Camp programs in Alabama are operated by the U.S. Space & Rocket Center and the Alabama Space Science Exhibit Commission. California and Florida locations are owned and operated by the U.S. Space Camp Foundation, a nonprofit organization. Alabama and Florida locations are accredited by the American Camping Association, and the new California locations are in the process of accreditation. For more info, check out www.spacecamp.com.

Astronaut Hall of Fame

The United States Astronaut Hall of Fame, dedicated to the Mercury, Gemini, Apollo, and now space shuttle astronauts, opened in early March 1990. It is operated jointly by the Mercury Seven Foundation (created by the seven original astronauts to encourage academic and technological excellence through college scholarships) and the U.S. Space Camp Foundation, and is located adjacent to the Space Camp in Titusville.

The Hall of Fame features personal mementos and equipment donated by the first astronauts, and showcases each Mercury and Gemini astronaut with exhibits and video. The facility showcases each of the 7 Mercury astronauts and 13 Gemini astronauts, who pioneered spaceflights in the 1960s.

The complex includes a presentation about the possible future of space travel, including exhibits on a space station and possible return trips to the Moon. A highlight is the "Shuttle to Tomorrow," a full-scale space shuttle *orbiter* featuring a multimedia theater in its cargo bay. It was added to the Hall of Fame in 1992.

The Hall of Fame is open daily, except Christmas. (There is an admission charge.) Follow Kennedy Space Center signs to Florida 405 and U.S. 1. Go one block east of the intersection toward the entrance to the Kennedy Space Center. Turn right for the Hall of Fame and Space Camp.

Space Talk

The **orbiter** is the reusable, winged portion of the space shuttle system, which actually orbits the Earth. The booster rockets fall into the ocean soon after lift-off, and the huge external tank breaks up when it enters the atmosphere over the Indian or Pacific Oceans.

Cosmic Facts

NASA doesn't endorse hotels, so I won't either. If you want information regarding your many options for fine accommodations during your visit to KSC, Cape Canaveral, or the Astronaut Hall of Fame, visit the Cocoa Beach Area Chamber of Commerce at www.cocoabeachchamber.com.

Cape Canaveral

Although people still think of the Kennedy Space Center—where the shuttle is launched and (most of the time) returns to Earth—as "Cape Canaveral," KSC actually is built on Merritt Island, which is across the Banana River to the west of the Cape. The actual Cape these days is the home of the Cape Canaveral Air Force Station, a military base operated by the Air Force's 45th Space Wing. Cape Canaveral was the site of the earliest American rocket launches, long before NASA's birth. It saw the first U.S. satellite launch, and our first manned space launches. Today, expendable military and commercial boosters and spacecraft are launched from this location.

The Air Force has used the Cape as their rocket test site since 1950. On July 24 of that year, a modified German V-2 rocket known as Bumper-8 became the first rocket to be launched from Cape Canaveral. In 1958, *Explorer 1*, our first satellite, and later the Mercury and Gemini missions were launched from the Cape. Today, NASA still owns and operates several launch facilities on the land barrier east of the Banana River known as Cape Canaveral.

Older readers may remember that, seven days after President John F. Kennedy was assassinated in 1963, the geographical feature known as Cape Canaveral had its name changed to Cape Kennedy. In 1973, however, public sentiment in Central Florida prompted the U.S. Congress to change the name back to Canaveral. The John F. Kennedy Space Center, as you've learned, is actually on Merritt Island, west and north of the Cape.

Johnson Space Center

The Lyndon B. Johnson Space Center (JSC) in Houston, Texas—you know, "Houston Mission Control"—was established in September 1961 as NASA's primary center for:

◆ Design, development, and testing of spacecraft and associated systems for human flight

◆ Selection and training of astronauts

◆ Planning and conducting human space flight missions

◆ Extensive participation in the medical, engineering, and scientific experiments carried aboard space flights

The center is adjacent to Clear Lake at 2101 NASA Road 1, about 25 miles southeast of downtown Houston, Texas, via Interstate 45. Additional facilities are located at nearby Ellington Field, seven miles north of the center. It, too, offers much to the visitor.

During space shuttle launches, KSC is in charge of the flight until the spacecraft clears the launch tower. At that moment, control of the mission shifts to JSC, and Mission Control.

The official visitor center of NASA's Johnson Space Center is Space Center Houston, a museum owned and operated by the Manned Space Flight Education Foundation, Inc. It is not federally funded. At the center, you can learn about and celebrate NASA's accomplishments. Everything you see in the facility is realistic. The designers kept in close contact with NASA to ensure the most accurate experience.

At Space Center Houston, you can see Mercury, Gemini, and Apollo spacecraft, touch and see moon rocks returned by the Apollo astronauts, and see exciting films on an IMAX screen. Land the shuttle in a simulator, and try your hand at rescuing a satellite. Get an up-to-the-minute summary of training and mission activities at the Mission Status Center, and take a tram tour of JSC's astronaut training facilities. Rotating exhibits in the center's main plaza guarantee an educational and entertaining visit. Outside, see an actual Saturn V moon rocket, a Mercury-Redstone booster, and an Apollo-era Little Joe rocket. The average visitor spends five hours at Space Center Houston.

The Center also offers bus tours to NASA's Sonny Carter Training Facility, the site of the huge spacewalk training facility called the Neutral Buoyancy Lab, or NBL. Named after veteran shuttle astronaut Sonny Carter, who died in a commuter plane crash in 1991, the facility houses the NBL, one of the largest swimming pools in the world. It's 202 feet long, 102 feet wide, and 40 feet deep and contains 6.2 million gallons of water. The NBL permits astronauts in spacesuits to work underwater on realistic mock-ups of the space shuttle and space station. Using weights to make their suits neutrally buoyant, astronauts get a close approximation of how it feels to work in free-fall. I'm one of the NBL's most frequent customers, and I can tell you that the feeling of a real spacewalk is very close to that offered by the NBL—except that in space, there are no air bubbles drifting upward, and the scenery is far better!

Space Center Houston is located right next door to JSC, at 1601 NASA Road, Houston, Texas 77058. That's about 25 miles south of downtown Houston in the NASA/Clear Lake area.

At nearby Ellington Field, a former Air Force base, there are no organized tours, but you can glimpse T-38 astronaut training jets roaring around the traffic pattern, and glimpse the famous "Vomit Comet" zero-g training aircraft, and the outsized Super Guppy transport plane.

Ames Research Center

You don't have to go to Florida or Texas to visit NASA. We're in northern California as well and have been ever since NASA's birth almost 45 years ago.

NASA Ames Research Center is located at Moffett Field, California, in the heart of Silicon Valley. Ames was founded December 20, 1939, as an aircraft research laboratory by the National Advisory Committee on Aeronautics (NACA) and in 1958 became part of

NASA. Ames specializes in aeronautics, astronautics, planetary astronomy, and life sciences research. This is where we create new knowledge and new technologies to help all NASA's endeavors. Ames is where we send space shuttle pilots to train for handling landing emergencies, where the tricky-to-handle orbiter is just a few feet above the runway. After dozens of runs in Ames's Vertical Motion Simulator, a blown tire on touchdown or a stiff crosswind will seem old hat to shuttle crews.

Ames is located on the border of the cities of Mountain View and Sunnyvale, at the southern end of San Francisco Bay. Ames occupies about 430 acres of land, and serves as host to a number of other federal, civilian, and military resident agencies on the adjoining 1,500-acre former naval air station, now known as Moffett Federal Airfield. You can't miss the huge old dirigible hangars on the field, which predate World War II.

In December 1987, the Ames Visitor Center opened at its present site, in a historic research building. Our visitors center is located in the building formerly used to test space capsules and heat shields, to see how they would react under what we call *hypervelocity*, the high speeds experienced during re-entry.

The Visitor Center is an information resource designed to educate the general public while showcasing a selection of Ames's world-class research and technology developments. It features a NASA U-2 high-altitude research aircraft.

To reach the Ames Visitor Center, take the Moffett Field exit from Highway 101, turn left in front of the Moffett Field Main Gate, and follow the signs to the Visitor Center.

> **Cosmic Facts**
>
> If you want to contact NASA's California office, here's the address and Web site:
>
> NASA Ames Research Center
> Moffett Field, CA 94035
> www.arc.nasa.gov

> **Space Talk**
>
> **Hypervelocity** is NASA's term for really, really fast. Ames engineers had to deal with the heating problems faced when returning from orbit, when humans would be traveling faster—five miles a second—than they ever had before.

> **Cosmic Facts**
>
> To find out about places to stay when visiting Ames, see the San Jose Silicon Valley Chamber of Commerce Web site at www.sjchamber.com.

Marshall Space Flight Center

In addition to the space centers already mentioned, you might want to visit the Marshall Space Flight Center in Huntsville, Alabama.

The visitor center for Marshall is the Alabama Space and Rocket Center, a dazzling collection of rockets and artifacts that documents America's early rocket advances. After

World War II, the Army's rocket team under Wernher von Braun moved into the Redstone Arsenal here and developed the first U.S. medium- and long-range ballistic missiles. Here, von Braun's team later developed the Saturn series of boosters that took America to the Moon. At the center, you can see huge test structures for rocket engine run-ups, and the restored test stand for the first tests of the Redstone booster, the first U.S. missile developed by von Braun's group.

Cosmic Facts

For more information on the Marshall Space Flight Center, check out the Web site: http://www.msfc.nasa.gov.

The Rocket Center has an excellent museum, housing real Gemini and Apollo spacecraft, mock-ups of the Skylab space station, and an IMAX theater. But the real highlight is the outdoor rocket park, dominated by a full-scale, 363-foot mockup of a Saturn V moon rocket, standing vertically above the center. The Saturn V stands surrounded by a Saturn I-B, Atlas and Redstone boosters, and an actual Saturn V lying on its side. Out front is an SR-71 Blackbird, the famous Mach-3 reconnaissance plane.

The Marshall Space Flight Center and the Alabama Space and Rocket Center are two of my favorite space destinations. They make it well worth spending a day, steeped in the history of rocket science.

Jet Propulsion Laboratory

The Jet Propulsion Laboratory (JPL) is NASA's lead center for robotic (that is, non-manned) exploration of the solar system. JPL is located on a 177-acre site at the foot of the San Gabriel Mountains near Pasadena, California, 12 miles northeast of Los Angeles. JPL is managed for NASA by the California Institute of Technology.

Cosmic Facts

The address for the Jet Propulsion Laboratory is:

Jet Propulsion Laboratory
4800 Oak Grove Drive
Pasadena, California 91109

Spacecraft developed by JPL have visited all the planets in our solar system except Pluto. JPL cameras and sensors are aboard satellites circling Earth to study the ozone, oceans, and other Earth phenomena. JPL will be in the thick of our efforts to explore Mars, the asteroids, and other planets, as we endeavor to understand how our solar system was formed.

To take a tour of JPL, you have to make an advance reservation at JPL's Public Services Office. Tours are available both for groups and individuals. The tour includes a multimedia presentation entitled "Welcome to Outer Space," which functions as an overview of JPL's history and accomplishments. Guests may also visit the Von Karman Visitor Center, the Space Flight Operations Facility, and the Spacecraft Assembly Facility, or SAF. In the

Cosmic Facts

For JPL reservations, contact:

Public Services Office
Mail Stop 186-113
Jet Propulsion Laboratory
4800 Oak Grove Drive
Pasadena, CA 91109

latter building, you can glimpse spacecraft under final assembly. The SAF is where I watched our Space Radar Laboratory take shape back in 1992–1994. The tour involves a lot of hiking, including a considerable amount of stair climbing. Wheelchairs can be accommodated with advance notice. There are no restrictions on photography while on the tour, so feel free to bring your camera. Dress for the weather. Tours proceed rain or shine.

Goddard Space Flight Center

The Robert H. Goddard Space Flight Center, named after the innovative professor, has been NASA's space science laboratory since it opened in 1959. It is located in Greenbelt, Maryland, not far from Washington, D.C. For more info, check out www.gsfc.nasa.gov.

Stennis Space Center

The John C. Stennis Space Center in south Mississippi is where NASA does its rocket propulsion testing. It is located on I-10 at Exit 2, 48 miles west of Biloxi and 45 miles east of New Orleans. For more info visit www.ssc.nasa.gov.

Hopefully one of these NASA centers is in your neck of the woods and you can visit soon. You'll find it both fun and educational.

The Least You Need to Know

♦ Your tax dollars paid for NASA facilities. NASA wants you to see what we're up to.

♦ NASA's various visitors' centers are must-see, and a great place for families.

♦ You need not limit your visits to KSC and JSC—NASA centers are found around the country.

♦ Call or check the Web in advance for information on visitor access or reservations at NASA centers like JPL.

Part 2

The Early Days

In Part 2, we'll look back at the earliest days of the U.S. space program from the end of World War II until the Soviets put the first man in space. We'll examine the origins of the U.S. space program, and recall the frustration that many Americans felt as the Soviet Union got off to a big early lead in the space race.

The uncoordinated and failure-prone U.S. efforts after *Sputnik* led directly to the formation of NASA. Fueled by more funding and presidential backing, NASA took over, and things began to improve dramatically.

We'll see how the first American satellites finally made it into orbit. Many of the early American satellites were scientific, weather, or communication satellites, whose observations from space have revolutionized the world.

Chapter 6

Wernher von Braun and Early U.S. Rockets

In This Chapter

- The V-2
- Missile specifications
- Space fever
- Cold War

People have always looked up at the night sky, with its Moon and its stars, with wonder. With our horizons widened by the telescope, we've always wanted to explore the universe and visit other planets, but only during the last half-century have we had the technology necessary to carry out this dream.

Today, humans have walked on the Moon. We've landed craft on Mars and Venus, and the Earth is orbited by thousands of satellites that have revolutionized the fields of meteorology, communications, and our understanding of the cosmos.

Remember, it was only a few generations ago that we began to make strides, spurred on by wartime advances, in space technology. In this chapter, we'll

Cosmic Facts

Here's a quick refresher course for those who might not have been paying attention in astronomy class: The Sun is the center of our solar system. The Earth is one of nine planets circling the Sun. Earth is the third planet from the Sun. The planets are, from nearest to farthest from the Sun: Mercury, Venus, Earth, Mars, Jupiter, Saturn, Uranus, Neptune, and Pluto.

take a look at the earliest days of modern rocketry, with a special focus on several of the innovators whose concepts and inventions led humankind into the Space Age. First and foremost among those movers and shakers is a man named Wernher von Braun.

Son of a Baron

Wernher von Braun was the son of a baron, and clearly a genius. As an engineering student, it took him only 18 months to earn his doctorate degree, and he was appointed the top civilian specialist in the German rocket program at the age of 20.

First Nazi Rocket

The Treaty of Versailles—the peace treaty that ended World War I—severely limited German military weapons development following the war. Its artillery experts looked around for technology that would supplement their conventional cannon. A German amateur rocket group in Berlin—of which von Braun was a member—came under German army sponsorship, and its talented core began to work for the Nazi establishment.

Early rocket pioneer Dr. Wernher von Braun came from Germany after World War II and became NASA's Deputy Associate Director.

(NASA)

After a series of experiments with small, liquid-fuel rockets, the German team came up with plans for a long-range missile carrying a large explosive *warhead*. This rocket, termed the A-4, later became known around the world as the V-2.

Space Talk

A **warhead** is a bomb attached to the nose of a rocket.

Dornberger's Role

The project manager for the A-4 was Walter Dornberger, and his technical director was Wernher von Braun. Six years before the start of World War II, Dornberger had been assigned to develop a rocket that could travel farther than the maximum range of any German gun and carry a payload (bomb) larger than anything then available in the German artillery. Under strict secrecy, the first launch of the A-4 came in June 1942 at Peenemunde on Germany's Baltic coast.

These first two tests failed, but on the third try the rocket traveled 118 miles. The A-4 rockets were not consistently successful until the spring of 1943. By that time the rockets were arcing high above the Baltic Sea, traveling at five times the speed of sound, and when they were aimed at England, there could be no defense against them.

Armed with a one-ton, high-explosive warhead, the rockets were first used as weapons in September 1944 (too late to alter the outcome of the war). By then Hitler had given the missile the name "Vengeance Weapon Number Two," or V-2 for short. Thousands were fired at London, Antwerp, and the Allied port facilities in France and Holland.

While involved in the Nazi war effort, neither von Braun nor Dornberger were allowed to do any planning for the use of rockets in space exploration. They kept those long-held ambitions to themselves. The sole object of their rocketry efforts, through 1945, was to deliver as many warheads as possible onto populated areas of England. The attacks killed thousands.

How von Braun's Efforts Helped the Allies

Since the missiles had only limited accuracy, many missed their intended targets in populated areas. The V-2 was a failure at both its terror and military missions.

It wasn't until the atomic bomb was married to the concept of a long-range missile that the world's view of the V-2 as a new kind of weapon system changed.

Cosmic Facts

Some 3,225 V-2s were launched in anger, each carrying more than a ton of explosive.

Because Hitler put so much time, money, and energy into developing a weapon that, in the long run, lacked the ability to alter the war, von Braun's efforts probably contributed unintentionally to an Allied victory. The V-2 effort was the German equivalent of our Manhattan Project (the code name for our project to develop an atomic bomb).

If those same resources had been applied to deploying the new German jet fighters, the war would have been prolonged, perhaps for years—or until the development of the atomic bomb. For the Allies were developing their own super weapon that did deliver on its promise to decisively bring an end to the conflict.

Operation *Paperclip*

Following World War II, von Braun and Dornberger saw the United States as the only country that could possibly support them in their ambition to continue development of rockets for space exploration. Their decision to surrender to the Allies, and not the Russians, would be instrumental in developing American rocket science during the Cold War.

The United States, anxious to give its own rocket programs a boost, sent teams into Europe behind the advancing Allies to capture the V-2, its developers, its technology, and the experts on von Braun's staff. This was part of a larger move to recruit German aviation scientists, engineers, and intelligence experts into the Allied cause, in view of the upcoming Cold War against the Soviet Union.

Cosmic Facts

One immediate improvement over the V-2 was in the building materials. Whereas the German rocket was made of heavy steel, the American rocket was built of the much lighter aluminum. When fully fueled, the Viking weighed five tons, about a third of the V-2's weight.

Von Braun and his rocket group slipped westward toward the front lines, and finally managed to find a U.S. Army unit that would accept his surrender.

Along with von Braun and his team, the United States also took as many V-2s and parts as it could find and hustled them back to the U.S., in an effort termed Operation *Paperclip*. Von Braun and most of his group wound up in far west Texas at Ft. Bliss, an Army artillery range.

The U.S. space program began with test launchings of those rockets from the White Sands Proving Ground in New Mexico, a few miles from Ft. Bliss. Walter Dornberger came to the U.S. and went to work for Bell Aircraft.

The Soviets were not idle, and had the same idea about capturing V-2 technology. They, too, recruited German rocket scientists to help their missile programs.

Viking and Redstone: First U.S. Efforts

While von Braun and his group got settled out West, the U.S. pressed ahead with rocket development. The first practical liquid-fuel rocket that the United States developed on its own was the Viking, a product of the Naval Research Laboratory and scientist Milton Rosen. The initial goal was to build a rocket that could carry 500 pounds to a 100-mile altitude.

There were 12 tests of the Viking rocket. The first 3 were all marred by premature cutoffs of the engines. Of the remaining 9, 7 were complete successes. One Viking that was launched in May 1954 lifted an 825-pound payload to an altitude of 158 miles.

First Space Pix

Another Viking feat was the return of the first space photography. The payload took pictures and then dropped the exposed film, protected in heavy canisters, back through the Earth's atmosphere for retrieval on the ground.

In 1954, Viking revealed the potential of space as a vantage point for weather prediction by photographing a tropical storm over south Texas.

Dr. Jones's Corner
The captured V-2s gave the U.S. program a jump-start. For example, the V-2 served as the bottom half of the first two-stage rocket. In February 1949, the V-2, coupled with a WAC Corporal rocket, reached an altitude of 244 miles, a record that stood until 1956.

Von Braun's own project, a follow-up to the V-2, was the Redstone rocket. Using a hand-me-down U.S. Air Force engine, the rocket could carry a nuclear warhead downrange 200 miles. Although the Redstone rocket would quickly be passed in power and range by other U.S. missiles, the Redstone was reliable, and earned its place in history. It never carried a nuclear missile downrange, of course, but it was used to boost the first American man into space (in 1961)—a milestone we'll look at in greater detail in Chapter 13, "Near Perfection."

Jupiter-C and Atlas

After Redstone, von Braun began work on his first multistage rocket, which he called Jupiter-C. He first tested the new rocket on September 20, 1956. The rocket broke all existing performance records. It flew to an altitude of 682 miles, breaking the record set by the V-2/WAC Corporal combination in 1949. It also flew 3,355 miles downrange.

Sixteen months later this rocket would be responsible for putting the first United States satellite successfully into orbit.

Space Talk

An **intercontinental ballistic missile** is a rocket capable of carrying a warhead from one continent to another.

Cosmic Facts

The Atlas rocket was far larger than any other booster that had been produced by the United States up until that time. It stood 75 feet tall, weighed 260,000 pounds, and produced 360,000 pounds of thrust at lift-off.

The rocket that would eventually send the first Mercury astronaut into Earth orbit went into development in August 1951. It was being developed by Convair, whose parent company was known as the Atlas Corporation. The new rocket would be known as Atlas.

Although expensive and clumsy as a weapon, Atlas did have a downrange capability of 5,000 miles, which meant that, for the first time, an *intercontinental ballistic missile* (ICBM) was feasible.

Up until 1955, nearly all the U.S. efforts in rocketry were driven by military requirements. Only a few sounding rockets, which measured atmospheric conditions at various altitudes, had been built for research. The Atlas, coupled with a variety of upper stages, later became a workhorse for NASA scientific and private commercial launches.

The Atlas is still flying today. The Atlas/Centaur rocket that first became operational in 1966 evolved into today's Atlas I and Atlas II. The Atlas/Centaur was the launch vehicle for *Surveyor 1*, the first U.S. spacecraft to soft-land on the Moon. Today, these rockets are made by Lockheed-Martin, which uses them for defense and commercial payloads launched from Cape Canaveral.

The Atlas I rocket is strong enough to put a 6½-ton satellite into low earth orbit, a 2½-ton satellite into a geostationary orbit (where the satellite appears to hover over a single region of Earth; see Chapter 10, "Early Communications and Weather Satellites"), or a 1½-ton satellite on a mission beyond Earth orbit.

Cosmic Facts

Other spacecraft launched by NASA Atlas/Centaurs include:

- The Orbiting Astronomical Observatories
- Applications Technology Satellites
- Intelsat IV, IV-A, and V series of communications satellites
- Mariner Mars orbiters
- *Mariner 10*, which made a flyby of Venus and three of Mercury
- *Pioneers 10* and *11*, which accomplished flybys of Jupiter and Saturn
- *Pioneer Venus*, a spacecraft that orbited Venus and sent several probes plunging through its atmosphere to the surface

The Atlas II stands 150 feet high. The first stage, at lift-off, produces 468,500 pounds of thrust. The Centaur stage has the same thrust as on the Atlas I, but carries 6,600 pounds more propellant, and so can burn longer to achieve a higher velocity.

An Atlas II can place a 14,950-pound payload into low Earth orbit, a 6,100-pound satellite into geostationary orbit, or a 4,270-pound satellite on a mission elsewhere in the solar system.

Titan and Thor

While Atlas was still in development, a parallel program was started for a rocket that would be even larger, able to go farther while carrying more. Its name was Titan, and when perfected, it would not only be a successful ICBM, but responsible for launching all the Gemini astronauts.

Thor was an Air Force intermediate range ballistic missile developed in the mid-1950s. When topped with a powerful second stage like the Agena rocket, this booster launched a series of military reconnaissance satellites, scientific probes, and some of the first U.S. lunar probes. The Thor booster later served as the core for the reliable Delta booster of today.

The Titan-Centaur

The original two-stage Titan missile soon grew into a powerful space booster. Called the Titan III-E, the rocket sandwiched the original liquid-fuel stages between two large solid-fuel rockets.

In 1974, NASA took the Titan III-E up a notch, launching the first of seven vehicles that topped the Titan core with the powerful Centaur upper stage, replacing the old Air Force third stage. The new Titan-Centaur became the booster of choice for deep space and planetary missions.

At lift-off the two solids, burning alone, produced 2.4 million pounds of thrust; the liquid propellant first stage ignited when the solids burned out, followed by the second stage. Then the Centaur, which produced 30,000 pounds of thrust from two main engines, ignited and burned for up to 7½ minutes. The Centaur could also restart in orbit, pushing robot spacecraft like Viking off to the planets.

The Titan-Centaur had an overall height of 160 feet. It bridged the gap between the last Saturn launches and the advent of the space shuttle. During that time it successfully launched six large interplanetary probes.

Titans Today

Titan rockets, like the Atlas, have never gone away. Today Lockheed-Martin, builder of the Titan family of vehicles, launches the Titan IV rocket both from Cape Canaveral and from Vandenberg Air Force Base in California. The primary customer for the Titan is the U.S. Air Force, but since the powerful Titan can lift the heaviest payloads, they are also used for commercial and NASA scientific missions.

This vehicle is 204 feet in height. It is capable of placing 39,000 pounds into an east-west orbit from the Cape—a payload comparable to that of the space shuttle.

Media Starts Space Craze

During the early 1950s, the idea of space travel became very popular in the United States, not because conquering space would give us a better chance to win the Cold War, but because of its connection with the human need to explore, to find out what was out there.

Arthur C. Clarke published *The Exploration of Space* in 1951. Walt Disney made a documentary called *Man in Space* in 1954. Science fiction movies featuring manned flights to the Moon or Mars became popular Hollywood themes.

Cuban Missile Crisis

The new Cold War rockets came perilously close to being used in the Cuban Missile Crisis in 1962. In October 1962, Soviet Premier Nikita Khrushchev, lacking a capable long-range missile force, put medium-range missiles in Communist Cuba, only 90 miles from Florida. After President Kennedy challenged this move and imposed a naval blockade on Cuba, both countries brought their missiles to a state of full alert.

It seemed for several days that an ICBM exchange using nuclear weapons was imminent. With a U.S. invasion of Cuba looming, Khrushchev backed down and removed the missiles from Cuba in exchange for an agreement that the U.S. would not forcefully remove Fidel Castro and his Communist regime from Cuba.

The Cold War was also "fought" in space. Even as both sides struggled to build a rocket that could deliver a hydrogen bomb accurately to a target on the other side of the world, the space race was on. At first the goal was an Earth satellite, but the finish line soon moved out to the Moon.

Reaching for Doomsday: Delivering an H-Bomb Reliably to a Target

The push to construct a practical and reliable ICBM led to a rapid growth in all facets of rocket technology. The very advances that would make possible an accurate bomb-delivering missile were the same needed to begin the exploration of space: reliable rocket engines, accurate guidance systems, and light-weight structures and materials. Because of the smaller size of the American hydrogen bomb, the U.S. was able to get by with smaller boosters for delivery of its nuclear weapons. The heavy Russian H-bombs produced an impressive rocket payload capability.

> **Dr. Jones's Corner**
>
> The U.S.'s Atlas rocket could put barely two tons into orbit, half that of the Russian R-7 missile.

Threat of Nuclear War

The existence of ICBMs meant that a surprise attack could be mounted on an adversary on the other side of the planet with little or no warning. With worries of a "first strike" by the Russians, America scrambled for a means of early warning and for a way to detect signs that Russia might be preparing such a strike. The vantage point of space held the key to both problems.

Controlling the "High Ground": Control of Space As Military Advantage

By the early 1960s, the U.S. could no longer fly over enemies like Russia to conduct strategic aircraft reconnaissance. The 1960 downing of a U-2 spy plane over Russia cut off direct intelligence of what the Soviets were up to. President Eisenhower gave the go-ahead to develop a satellite system, called Corona, that could photograph Russian rocket and bomber bases. This combination of booster capacity and orbital photography was the same technology needed to begin the exploration of the Moon and planets.

Reliable, powerful boosters also gave the U.S. the ability to put early-warning satellites into orbit. Looking down on Russian bomber and missile bases, these satellites could detect the hot flash of a missile launch, and give about 30 minutes of warning to the President and military forces. Offering such a vantage point, space quickly became an arena busy with U.S. and Russian military activity.

On the American side, this vigorous pursuit of reliable access to space in the late 1950s and early 1960s led to a stable of rocket boosters that gave NASA the means to build the rockets and spacecraft needed to fulfill its mission.

The Least You Need to Know

- ◆ The Nazis developed the first practical guided missile, the V-2 rocket. Wernher von Braun, who dreamed of space travel, first developed the deadly terror weapon for Germany.
- ◆ While both Russia and America captured German missile technicians and technology, the Americans recruited von Braun, whose talents jump-started the U.S. rocket development program.
- ◆ The V-2's potential led both Russia and the U.S. to marry the guided missile to the atomic bomb, leading to production of Intercontinental Ballistic Missiles (ICBMs).
- ◆ Both Cold War adversaries modified their ICBMs from military use to provide the first crop of space boosters, and their descendants are still in use today.

Chapter 7

Russia First Out of the Space Gate

In This Chapter

- ◆ Sputno-phobia!
- ◆ Dog in space
- ◆ Ike's panel
- ◆ NASA is born

After the Berlin Airlift in 1949 and the Korean War from 1950–53, the Cold War steadily squeezed the world in the grip of fear. After the Soviet Union exploded its first atomic bomb in 1949, the United States military knew the USSR was working on a missile system sophisticated enough to deliver those warheads to North American soil.

So it was with considerable discomfort that U.S. leaders learned during the summer of 1957 that U.S. radar had picked up evidence of Soviet intercontinental missile testing.

The testing had actually begun the previous spring, on May 15, 1957, when the Soviets attempted to launch their R-7 rocket. This initial test lasted less than two minutes. After lift-off, a ruptured fuel line caused an explosion that destroyed the rocket.

A second test on June 9 was also a disaster. The rocket simply failed to fire—quite an anticlimax following a dramatic countdown.

The next try came on July 11, and this time the rocket got off the ground. Unfortunately, the guidance system wasn't working and the rocket quickly lost control and blew up.

Then, on August 21, success. The R-7 launch went perfectly and the rocket ended up crashing back to Earth a record-breaking 4,000 miles from the launch site. For the first time, Korolev's team had flown a missile that could realistically travel from the Soviet Union to the United States.

Cosmic Facts

As it turned out, the June 9 fizzle of Sergei Korolev's R-7 came about because someone had installed a fuel valve upside down.

Space Talk

Sputnik is Russian for "companion," as in "companion of the Earth."

A second successful test followed in September. The Russians decided to reveal their missile prowess by putting an artificial satellite into orbit—and the West quickly realized that the same rocket could easily hurl a hydrogen bomb to America.

Korolev's team grappled with what to use as a payload on the next R-7. Some Soviet scientists had wanted to put a heavy satellite into orbit, one that contained scientific experiments—but the need to be first pushed the Soviets toward a simpler plan. (The Americans had announced they were planning to launch a satellite late in 1957.) The first artificial satellite, the Russians decided, was to be a 184-pound silver ball named *Sputnik*.

Sputnik: Starting Gun of the Space Race

For *Sputnik*, a special nosecone was built to fit the tip of the R-7 rocket. Inside the nosecone was the satellite. The launch took place on the evening of October 4, 1957.

Cosmic Facts

The day after *Sputnik* went into orbit, the headline on the front page of *The New York Times* read: "SOVIET FIRES EARTH SATELLITE INTO SPACE; IT IS CIRCLING THE GLOBE AT 18,000 MPH; SPHERE TRACKED IN 4 CROSSINGS OVER U.S."

Once again the R-7 worked perfectly, pushing the satellite up and out of Earth's atmosphere and into orbit.

Americans remember the shock of *Sputnik*'s launching—and the fear. To give up space to the Soviets was to give up the world to Communism.

The race was on.

K9 Adventure

The next step for the Soviet space program was to put a living creature into space. They had been putting dogs into high-altitude rockets for years to test their canine reactions, so the Soviets decided to put a dog into orbit using a larger *capsule*, complete with life support systems, food, water, and monitoring equipment.

I can't say the dog, Laika, a mixed-breed terrier, was lucky to be the first animal to orbit Earth, because pressured Russian engineers had not had the time to design a return system for the capsule. Laika was killed by an injection of poison a week after launch.

NASA Is Born

With the Russians in space, scientific experts in the U.S. began to suggest that America should form a "National Space Establishment" to oversee the U.S. Space program.

These rumblings came from men who believed that rockets could be used for more than just immediate military advantage. One of the organizations that endorsed the formation of such an establishment was the American Rocket Society, which you learned about in Chapter 3, "What Makes a Rocket Go Up?"

The society appointed a committee, the Space Flight Technical Committee, to analyze the practical necessities of a space exploration program. The committee's eventual report laid out a 25-year program which included orbiting spacecraft around the Earth, sending robotic satellites to explore other planets, and landing a man on the Moon.

The report also stressed the necessity of a civilian space agency to oversee the program. U.S. President Dwight Eisenhower's science advisors read the report and gave it a glowing endorsement on December 30, 1957.

Ike Appoints a Panel

A little more than a month later, on February 4, 1958, only days after the United States launched its first satellite into orbit (*Explorer I*—but more about that in Chapter 9,

Space Talk

Capsule was the original preferred term for the spacecraft that held the cargo, whether that cargo was alive or not. It wasn't until the American astronauts complained that they preferred the word spacecraft that the latter term was used, but the term "space capsule" never completely went away.

Dr. Jones's Corner

Some American experts had criticized the first *Sputnik* for being diminutive—only 184 pounds. After *Sputnik 2*, no one could criticize this canine-carrying spacecraft as little. It weighed more than half a ton, including Laika, and was accompanied into orbit by the empty booster stage, itself weighing about 8,000 pounds.

Cosmic Facts

James Doolittle, who had made headlines during World War II with his 1942 bombing raid of Tokyo, was General James Doolittle in 1958 and was a member of the presidential committee that formed NASA. That committee was chaired by Edward Purcell, a Nobel prize winner and noted Harvard physicist.

"Explorer and Pioneer"), Eisenhower appointed a panel, within the President's Science Advisory Committee, to form the civilian space agency that his advisors said was needed.

Of course, not everyone agreed with giving civilians control of the space program. Leaders of both the U.S. Army and Air Force strongly believed that space was a military domain, and that to turn the "exploration" of space over to a civilian organization was unwisely compromising the security of the nation.

Space was to be conquered and dominated, their thinking went. This disagreement lingered for decades, and scientists and generals still argue over how much to spend on civilian and military space programs.

"Introduction to Outer Space"

Eisenhower's panel released a report called "Introduction to Outer Space," and in it they named four reasons developing space technology was essential:

- Opportunities for scientific research and experimentation
- International prestige
- National defense
- Man's compelling urge to explore

While the report said that space offered great opportunities in reconnaissance (monitoring enemies and warning of attack), as well as in communications and meteorology, it also spoke of the disadvantages of developing space weapons—they would inevitably turn space into a war zone.

Congress Makes It So

In the end, Congress decided that the United States' extremely public space program was going to spread the word of peace and democracy. All civilian missions were to be peaceful and scientific. This said, there was an unspoken agreement that the military advantages of developing space technologies were very important and thus would continue out of the public eye.

The world, then as now, would have been suspicious if the Pentagon had taken sole control of the space program and used it to develop weapons that could be aimed either at Earth or other spacecraft. But there's a blurry line between our civilian and military programs even today. My two Space Radar Lab flights led to a shuttle radar mapping mission, STS-99, which produced an accurate topographic map of the world. While this map is scientific in character, the mission was funded by the National Imaging and Mapping Agency, which produces maps for the military. Space exploration has scientific, commercial, and yes, military benefits.

On April 2, 1958, President Eisenhower's space bill was submitted to Congress. Some minor changes were made during debate but the bill was passed and Eisenhower signed the act—P.L. 58-568, the National Aeronautics and Space Act of 1958—on July 29. The National Aeronautics and Space Administration opened its doors on October 1, 1958.

Cosmic Facts

U.S. Space Scorecard for 1958
Failures: 13
Successes: 5 (3 Explorers, 1 Vanguard, 1 Atlas)

National Advisory Committee for Aeronautics

When NASA was born, it absorbed into its structure another organization that had been heavily involved in the space program up until that point, the National Advisory Committee for Aeronautics (NACA). The Army's Jet Propulsion Laboratory and von Braun's rocket team at Redstone Arsenal, also became a part of NASA.

Cosmic Facts

The NACA never ran the space program, but it did have administrative responsibility over whole programs, including Vanguard (see Chapter 8, "Early Frustrations: Project Kaboom [A.K.A. Vanguard]").

The new NASA organization would become the world leader in space exploration one day, but at its birth, it was far behind the Russian space organization. There was a lot of catching up to do.

The Least You Need to Know

- Russia launched the first satellite, called *Sputnik*, on October 4, 1957.
- The first living creature in space was Laika, a Russian dog.
- NASA, by design, is a civilian rather than a military organization. All its research and mission results are available to the public.
- NASA's birthday is October 1, 1958.

Early Frustrations: Project Kaboom (A.K.A. Vanguard)

In This Chapter

◆ Making NASA necessary

◆ Many explosive failures

◆ Second U.S. satellite in orbit

◆ A positive legacy

When people look back at the many failures in the early history of the space race, they are most apt to focus on Vanguard, the Navy's rocket program of the mid-1950s. For years filmmakers used footage of Vanguard after Vanguard exploding, falling over, or pin-wheeling to disaster to illustrate the United States' early frustrations in rocketry.

You Can't Blame NASA

Vanguard's very public failures can't be blamed on NASA, because the project existed before the space agency was born. I include it here because it illustrates perfectly why NASA was necessary.

The early U.S. space program was fragmented and disorganized. The Army, Navy, and Air Force were all involved and were more often in competition rather than cooperation with one another.

"Designed by Rube Goldberg"

The Army, with Wernher von Braun leading the effort (see Chapter 6, "Wernher von Braun and Early U.S. Rockets," for more on von Braun), wanted to be the first to put a satellite into orbit, but their rocket was made up of the adapted parts or leftovers from other missile programs, prompting author William E. Burrows to write that the Army launcher looked to have been "designed by Rube Goldberg."

> ### Dr. Jones's Corner
>
> One nontechnical reason that influenced the choice of Vanguard as the launcher for the first U.S. satellite attempt was that it was designed by Americans, rather than the Germans, who'd contributed so much expertise to the other U.S. rocket options. This sentiment led to some bitter irony when, in the long run, Vanguard failed, and the German-designed Redstone rocket pushed the first U.S. satellite into orbit.

> ### Dr. Jones's Corner
>
> The Vanguard first stage was a descendant of the Viking rocket that, on May 24, 1954, successfully carried an 852-pound scientific payload from White Sands, New Mexico, to an altitude of 158 miles, considered to be beyond the threshold of space. Vanguard looked like a good bet at first.

The Jet Propulsion Laboratory had a rocket it wanted to use, too. But in the long run it was the Naval Research Laboratory's (NRL) proposal that won out. The NRL would adapt the very successful Viking research rocket to be the first stage of the new Vanguard booster.

Second Priority

The proposal for Project Vanguard called for launching a 40-pound satellite into orbit using the Viking first stage, an Aerobee second stage, and a new third-stage rocket. The Glenn L. Martin Company, the same firm that produced the Viking, was the prime Vanguard contractor.

The only problem was that the Martin Company was already under contract and busy making Titan ICBMs. The Vanguard program received second priority in everything, from inexperienced personnel to cramped, uncomfortable office space.

First Rocket Designed to Orbit Satellites

Vanguard has the honor of being the first American rocket program that had nothing to do with a weapons system. Its purpose was simple: put the first satellite into orbit.

Although a Vanguard rocket did not launch the first U.S. satellite, and the program was filled with disasters, many lessons were learned that would come in handy during later programs.

Vanguard's mission was to launch a series of *IGY* satellites. IGY stood for International Geophysical Year, which ran from July 1, 1957 through December 31, 1958. IGY plans called for a host of new scientific efforts aimed at understanding the Earth and its surroundings in space.

First Attempt Explodes on Launch Pad

Vanguard's designers thought they had the inside track on orbiting the first satellite. The first two tests, on December 8, 1956, and May 1, 1957, did not contain satellites, but produced good results. But before a third Vanguard test could be run, the Soviets stunned the world when they put their first Sputnik in orbit on October 4, 1957.

The third Vanguard test had been scheduled for later in the fall, but by this time the U.S. was in a hurry to get its own satellite into orbit, so the White House ordered the Navy to announce that the next Vanguard rocket would attempt to put a satellite into orbit.

This mission had been designated TV-3 (for Test Vehicle). The previous tests had flown only with live first stages, but on TV-3, all three stages would have to work together for the first time.

An Aluminum Sphere

The Vanguard rocket carried a tiny 3.25-pound, 6.4-inch-diameter spherical satellite made of aluminum. It didn't get very far—in front of a national TV audience, the rocket staggered only four feet into the air, sank back, and exploded on its Cape Canaveral launchpad.

Orange Flame and Dirty Black Smoke

With the world's press watching, U.S. hopes for quickly matching the Russians' satellite success vanished in a cloud of orange flame and dirty black smoke.

The knowledge gained from Vanguard's failures is given credit for the U.S.'s eventual success in

Cosmic Facts

This December 6, 1957, failure illustrated the big lead the Russians had in booster power. The Vanguard rocket had a first-stage thrust of 27,000 pounds. That was about three percent of the thrust of the Soviet R-7 rocket that had put the first two Sputniks into orbit.

Cosmic Facts

The British press had a field day with the embarrassing Vanguard explosion. The London *Daily Herald* called it "FLOPNIK!" while the London *Daily Express* called it "KAPUT-NIK!" Soviet premier Khrushchev was reported to have joked that the Americans should have named their rocket "Rearguard."

putting a satellite into orbit. But a Vanguard rocket was not the booster for the successful orbiting of *Explorer I* on January 31, 1958.

Instead, the Army Ballistic Missile Agency launched *Explorer I* aboard a four-stage version of the Redstone missile, a booster designed by Wernher von Braun's team. Von Braun, in fact, had succeeded in putting a satellite into orbit on his first try.

Vanguard soldiered on despite its embarrassing debut. But things did not get much better. A second failure occurred on February 5, 1958. Then, Vanguard's luck seemed to turn.

Finally, Success: March 17, 1958

The next Vanguard rocket carried the *Vanguard 1* satellite into orbit on March 17, 1958. This was the second U.S. satellite successfully placed in orbit.

Cosmic Facts

Vanguard 1 remains the oldest man-made object in space, expected to orbit Earth for about 1,000 years.

Like the grapefruit-sized payload carried aboard the failed TV-3, *Vanguard 1* was a 3.25-pound, 6.4-inch diameter aluminum tracking satellite.

Vanguard 1 was the first satellite to use solar cells so that it could create and supply its own power. On future missions, when more power was needed, these cells grew into *solar panels*.

The next four Vanguard attempts failed, but the Navy successfully launched *Vanguard 2* on February 17, 1959.

Space Talk

Solar panels are flat panels that protrude from satellites. They gather rays of the sun and convert them into electricity.

Technology Testbed

Vanguard's legacy is twofold. Films of the early days of space exploration will always include the disasters, fireball after fireball exploding over Cape Canaveral. But, on a far more positive note, the rocket's upper stages formed the basis for successful rocket combinations employed on Atlas-Able, Thor-Able, and Scout rockets.

Cosmic Facts

About Vanguard

Classification: Space Launch Vehicle

Length: 72 feet

Diameter: 3 feet, 9 inches

Date of First Launch: December 8, 1956

Date of Final Launch: September 18, 1959

Successes: 3 of 14 launches

Vanguard developed technology that was used by future programs. For example, a modified Vanguard upper stage formed the basis of the Atlas-Antares second stage. The Atlas-Antares was used in NASA's Project Fire to test proposed Apollo re-entry vehicle designs. Vanguard's bottom line was that it showed our determination to succeed in space, even in the face of failure, and to carry on in public, sharing our failures and successes with the entire world.

The Least You Need to Know

- ◆ The Vanguard project came along before NASA was born, and its failures underlined the fact that NASA was necessary to oversee and coordinate our disorganized space effort.

- ◆ Vanguard was the first rocket program without direct ties to a weapons program.

- ◆ Vanguard's many failures, some of them on live TV, overshadowed some solid achievements.

- ◆ *Vanguard 1* was the second U.S. satellite in orbit, launched on March 17, 1958.

Chapter 9

Explorer and Pioneer

In This Chapter

- ◆ U.S. in orbit
- ◆ Breaking free
- ◆ Visiting neighbors
- ◆ Exploring the solar system

As we close the pre-NASA portion of our space program history, I should tell you that not everything was a bust before NASA got involved. In this chapter, I'll tell the story of the Explorer and Pioneer programs, both of which *were* successful. These programs first put American hardware into orbit around the Earth, and then beyond, into the solar system.

U.S. in Orbit (Finally): *Explorer I*

The first U.S. satellite to achieve orbit was *Explorer I*, manufactured by the Army Ballistic Missile Agency and the Jet Propulsion Laboratory. It was put into orbit only after von Braun's team was turned loose for the attempt in early November. The Army team stacked two smaller, solid-fuel stages on top of the reliable Redstone booster, and on January 31, 1958, fired the stack from Cape Canaveral. This Jupiter C rocket carried the satellite, which was contained in the rocket's fourth stage; the satellite was 6¾ feet long and 6½ inches in diameter. Shaped like a slender dart, the back half contained the solid-fuel rocket, the front half the three instruments and two radio transmitters. Four wire antennae hung from the center of the satellite like whiskers.

Explorer I's mission was to take the measure of the space environment. It contained instruments to measure possible bombardment by tiny meteorites or energetic particles called cosmic rays, and thermometers to keep track of the temperature.

It was *Explorer I* that discovered the Van Allen radiation belts, which deflect and trap charged particles from the sun and the galaxy. The doughnut-shaped belts were named after the cosmic ray instrument scientist, Dr. James A. Van Allen.

The Pioneer Program

The first U.S. spacecraft to attempt to leave Earth's orbit and explore other parts of the solar system was known as Pioneer. The first three satellites were known as *Pioneer 0*, *Pioneer 1*, and *Pioneer 2*.

Each was designed to go into lunar orbit. As it turned out, none of the three satellites had a completely successful mission, but they provided valuable information to scientists nonetheless.

Bumps in the Road

The Air Force ran the first Pioneer mission. *Pioneer 0* blew up only 77 seconds into its flight, on August 17, 1958; the first stage of its Thor rocket exploded. The satellite had originally been called *Pioneer 1* but was (justifiably) changed to *Pioneer 0* after the failure.

After the failed attempt, NASA took over the project. In fact, *Pioneer 1* was the first launch ever made by NASA, taking place on October 11, 1958. Unfortunately, because of a programming error in the Thor rocket's upper stage, *Pioneer 1* lacked the speed to escape the Earth's gravitational field. It did, however, reach an altitude of 70,748 miles above the Earth, about a quarter of the way to the Moon, and provided data on the extent of the Earth's radiation belts. *Pioneer 1*'s launch vehicle re-entered the Earth's atmosphere over the Pacific Ocean two days later.

Pioneer 2, launched on November 8, 1958, failed when the third stage of its Thor rocket failed to ignite.

Pioneer 3: Still Can't Break Free

Pioneer 3, launched on December 6, 1958, was smaller than the first two had been and was to be blasted free of Earth's gravity atop a Juno 2 missile. The mission was a simple one: measure space radiation as it traveled to and orbited the Moon.

Once again, all did not go well. The launch vehicle's first stage stopped burning prematurely and the spacecraft failed to escape Earth's gravitational pull.

The vehicle did not make it to the Moon. It did, however, reach an altitude of 63,589 miles and discovered a second, previously unknown, radiation belt around the Earth. The craft then burned up as it re-entered Earth's atmosphere.

Cosmic Facts

Satellite Firsts

October 4, 1957: First satellite in space, *Sputnik 1*.

November 3, 1957: First animal in space, Laika the dog.

February 1, 1958: First U.S. satellite in space, *Explorer 1*.

April 12, 1961: First man in space, Yuri Gagarin.

June 16, 1963: First woman in space, Valentina Tereshkova.

August 19, 1964: First geostationary telecommunications satellite, *Syncom 3* (see Chapter 10).

April 6, 1984: First satellite repaired in orbit by the space shuttle.

November 16, 1984: First satellites returned from orbit by the space shuttle.

Near Miss of the Moon: *Pioneer 4*

Pioneer 4 was launched on March 3, 1959, and at last, all went well. It was the first U.S. spacecraft to venture outside Earth orbit. (The craft failed to beat the Soviets at this feat, as they had successfully launched *Luna 1* several weeks earlier.)

A camera trigger system was tested, to see if scientists could operate a camera by remote control from those distances. The system worked, but *Pioneer 4* itself contained no camera.

Cosmic Facts

Shaped like a cone, *Pioneer 4* was 1½ feet tall and ¾ foot in diameter at the base. It was coated with special paint designed to keep it cool under the relentless heat of the Sun, and contained two Geiger counters with which it measured radiation.

The flight plan called for the craft to pass within 18,600 miles of the Moon's surface, but *Pioneer 4* was a little off course, and it got only within 37,200 miles.

Pioneer 5: Maps Interplanetary Magnetic Field

Pioneer 5 was launched atop a Thor-Able rocket on March 11, 1960, and also left Earth's gravitation completely, bound for the furthest reaches of the solar system.

The satellite was a 2.1-foot sphere. Attached to it were four "solar paddles," which gathered up the rays of the Sun. The paddles enabled *Pioneer 5* to generate its own 16-watt power. It was designed to record the first map of the interplanetary magnetic field and returned signals to the earth for 106 days.

Cosmic Facts

When last heard from, *Pioneer 5* was more than 22 million miles away. It's still in an orbit around the Sun today.

Dr. Jones's Corner

The *Pioneer 6* through *9* series of satellites—each of which was about 3 feet in diameter—had solar panels so they could power their instruments and 79-watt transmitters. These space probes gave us our first look at space conditions between the planets, and their larger Pioneer cousins were first to reach—and pioneer—the giant outer planets, Jupiter and Saturn.

The Solar Wind Came Rolling In

The next three Pioneer missions—*6, 7,* and *8*—were designed to measure the solar wind, cosmic rays, and the solar magnetic field. They gave scientists their first report on space weather—weather conditions in space caused by the Sun.

The Pioneer missions have been called the most economical missions in NASA history, providing great bang for the buck. *Pioneer 6,* for example, in orbit around the Sun, is still operating 36 years after it was launched (though NASA no longer monitors it).

Pioneer 6 was launched December 16, 1965; *7* was launched August 17, 1966; *8* on December 13, 1967; and *9* on November 8, 1968. Another Pioneer mission, code-lettered *E,* was launched on August 27, 1969, but failed to reach orbit because of a Delta launch-vehicle failure. All these Pioneers were launched atop Delta E boosters, descendants of the Thor missile.

Jupiter-Bound

While another series of spacecraft focused on Venus and Mars, *Pioneer 10* was designed, to use a football term, to "go deep." Nine feet in diameter, *Pioneer 10* was launched March 3, 1972, atop an Atlas/Centaur rocket. It provided the first up-close photographs of Jupiter in December 1973. The spacecraft came closest to Jupiter on December 3, when it passed within 124,000 miles of the solar system's largest planet.

Acting as a scout for future Voyager missions, *Pioneer 10* also measured Jupiter's atmosphere, magnetic field, and moons, as well as the interplanetary magnetic and dust particle environment, the solar wind, and cosmic rays.

Pioneer 10 was not placed in orbit around the Sun as other interplanetary craft had been, but was rather set on a course to "escape" the solar system completely. Thus, *Pioneer 10* became the first human-made object to leave our solar system.

Saturn Has a Nice Ring to It

Launched on April 6, 1973, on another Atlas/ Centaur booster, *Pioneer 11* paid Jupiter a second visit—passing within 21,000 miles on December 4, 1974—then used Jupiter's gravity to sling-shot out at high velocity toward Saturn, where it would arrive five years later.

On September 1, 1979, *Pioneer 11* passed within 13,000 miles of the ringed planet. It then performed the same sort of experiments regarding Saturn's environment as *Pioneer 10* had for Jupiter.

> **Dr. Jones's Corner**
>
> Both *Pioneers 10* and *11* contain a gold plaque with etchings showing a man, a woman, and the location of our Sun and planet—just in case it should be found one day by extra-terrestrial life.

Pioneer 11's mission ended on September 30, 1995, when it had moved so far from the Sun that its nuclear generator (using heat from radioactive plutonium) could no longer produce sufficient power for it to transmit. *Pioneer 10* lasted longer, operating until 1997, and NASA still uses its faint signal to train flight controllers on future space probe operations.

Venus, I Love You

It was Venus's turn to have its picture taken when *Pioneer 12*, known as the Pioneer-Venus Orbiter, was launched by an Atlas/ Centaur rocket on May 20, 1978. On December 4 of that year, *Pioneer 12* was inserted into orbit around Venus.

Since Venus is always shrouded in thick clouds of sulfuric acid, ordinary cameras are useless for viewing the surface. Instead, *Pioneer 12* carried a *radar altimeter*.

> **Cosmic Facts**
>
> It took only a few minutes less than a full day for the Venus Orbiter to make one complete revolution around Venus. The orbit was an extreme ellipse, with an apogee (maximum distance from the planet) of 41,000 miles and a perigee (minimum distance) of only 93 miles.

Space Talk

A radar altimeter bounces radar signals off the terrain below, then measures the returning signals in order to make a relief map.

Once in orbit, the spacecraft made radar maps of the planet's surface, mapping 93 percent. It also constructed global maps of the planet's cloud formations, and performed analyses of Venus's ionosphere and atmosphere. And because of good fortune, *Pioneer 12* was in position to gather information about two comets.

Although the spacecraft was designed to operate for only eight months, it continued to transmit data until October 8, 1992, more than 14 years after it was put into space. At that time, the Venus Orbiter ran out of control propellant, so it was slowed down enough to fall into Venus's atmosphere, and burned up. The Orbiter carried 17 experiments and most of that equipment worked until the end.

Venus Multiprobe

Gathering data about Venus at the same time as the Venus Orbiter was its sister mission, *Pioneer 13*, the Venus Multiprobe. The Venus Multiprobe was propelled into space on August 8, 1978, by an Atlas/Centaur rocket. It did more than simply examine the planet from orbit as its predecessor was doing. *Pioneer 13* actually sent a cluster of four heat-shielded probes down into Venus's atmosphere. They were released from the main carrier spacecraft itself, powered by solar cells, in mid-November 1978.

Space Talk

Sunlight is so intense in space that satellites need to have special paint and insulation to reflect the heat. Some spacecraft even carry a thin, lightweight sunshade to keep the vital instruments cool. Any spacecraft headed down through an atmosphere also carries a **heat shield**, which protects the vehicle from the flaming heat of re-entry usually by charring and burning away. Heat shields are well-known features on piloted spacecraft.

The Probes

Pioneer 13's large probe had three parts: a sphere-shaped body, a protective cover, and a *heat shield*. It was one and a half meters in diameter. The protective covering and heat shield were jettisoned, and a parachute was deployed so that the probe—after an hour and a half of descent—could drift to the surface.

The smaller probes also had three parts, but the heat shield and protective cover remained attached to the sphere, which was about three feet in diameter—and there was no parachute. The smaller probes weighed about 165 pounds each.

The Surface of Venus

They entered the Venus atmosphere at 25,400 miles per hour, carrying equipment to measure the density, pressure, and temperature of Venus's atmosphere. One of the probes was large, while the other three were lighter, simpler versions of it.

The four probes went to different locations on Venus. Amazingly, one of the probes survived its landing and continued to broadcast signals for more than an hour before the heat and pressure at the surface cooked or crushed it into oblivion.

The main satellite also was targeted into the Venusian atmosphere, and it, too, was able to measure the density and composition of the atmosphere before it burned up on that same day.

Explorer I gave the U.S. its first taste of success in space, as it was not only placed in Earth orbit but made an important discovery (the Van Allen belts) as well. The Pioneer project taught us how to go beyond Earth orbit and visit our celestial neighbors throughout the solar system.

Cosmic Facts

Spacecraft headed for Venus, Mercury, or Mars are usually equipped with solar panels to produce the electricity they need. Those headed beyond Mars's orbit, where solar energy drops off dramatically, use nuclear thermoelectric generators to provide power.

The Least You Need to Know

◆ *Explorer I* was the first U.S. satellite to make it into Earth orbit.
◆ The launch of *Pioneer 1* was the first space probe to be supervised by NASA.
◆ The Pioneer series was the first to extend our reach out into the solar system.
◆ Like the Russian Venera landers, the *Pioneer 13* multiprobe returned signals from the surface of Venus.

Early Communications and Weather Satellites

In This Chapter

- ◆ Bouncing waves
- ◆ First communication satellites
- ◆ Receive and retransmit
- ◆ Weather satellites
- ◆ Saving lives

NASA has contributed to all fields of technology, but none more than the two areas of communications and meteorology. Let's talk about talking first.

Communications

We live in a world in which instant communications are taken for granted. You can pick up the phone and be chatting with someone in China in seconds. Go on the Internet and you can e-mail someone at the South Pole. Turn on the TV and you can see video of the Earth rolling by below the windows of the space shuttle or space station. But 50 years ago, things were very different.

The problem with communicating with someone on the other side of the world was that radio waves almost always travel in a straight line. You can't make them bend. The solution was to get them to change direction by bouncing them off something.

By placing a series of satellites in orbits around the earth, it was theorized, one could bounce a single radio or television signal from satellite to satellite, until it went all the way around the world.

SCORE

The first U.S. telecommunication satellite was launched in 1958. It was an Atlas missile called *SCORE* (Signal Communication by Orbiting Relay Experiment), and for the first time, voice communication was established via satellite.

The Atlas acted as a high-altitude broadcast antenna, carrying a tape recorder that ran a continuous tape loop. On the tape was the voice of U.S. President Dwight Eisenhower, who gave the world a Christmas greeting. *SCORE* lasted only 12 days, but it showed the potential of an orbital relay satellite.

Echo Program

NASA's first experimental communications satellite was called Echo. The Echo satellites (there were several) were large, lightweight metal balloons, 100 feet in diameter. Communications (television, radio, and telephone) signals were bounced off the Echo satellite to travel from one point on the Earth to another.

Echo 1, launched on May 13, 1960, never made it into orbit because its launch vehicle, a Delta rocket, failed. The first Echo to achieve orbit was technically *1A*, on August 12, 1960, but it's commonly referred to as *Echo 1*. *Echo 2* was put into orbit by a Thor Agena B rocket on January 25, 1964.

Echo 1A re-entered the Earth's atmosphere and burned up on May 24, 1968, followed by *Echo 2* on June 7, 1969.

Cosmic Facts

The most memorable thing about Echo was that, because of its size and highly reflective coating, it was easily visible to the naked eye. Nearly 40 years ago, many stargazers stood outside at night and watched the shooting star that was Echo streak across the night sky, faster than any airplane.

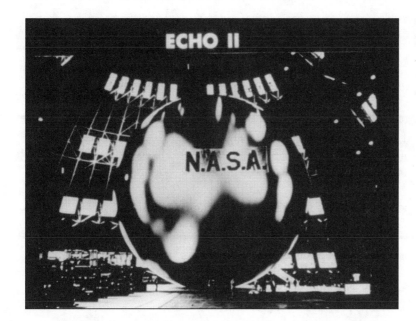

Echo 2 *was 30.5 meters in diameter. Communications signals were bounced off it to travel from one point on the Earth to another.*

(NASA)

Although Echo was passive—it merely served as a mirror that bounced signals back to Earth—another satellite put into orbit in 1960 did a little bit more. It was called *Courier*, and was launched by the Department of Defense. Like all communications satellites in use today, *Courier* received signals that were aimed at it from Earth and then retransmitted those signals back down to a ground station. It worked for only 17 days, the reason for its failure unknown, but the commercial and military value of its technology was clear. Retransmitting the signal, rather than bouncing it, meant that the signal was both stronger and more accurately aimed.

Syncom

The first three Syncom satellites were all launched by Delta-B rockets from Cape Canaveral in 1963 and 1964. They would bring to reality the dream of the visionary science fiction writer Arthur C. Clarke, who first suggested a new method of global communication.

Arthur C. Clarke In 1945, during the last days of World War II, a British Royal Air Force officer, also a member of the British Interplanetary Society, wrote a letter to a British technical journal called *Wireless World*.

The letter said: "… I would like to close by mentioning a possibility of the more remote future … perhaps half a century ahead. An 'artificial satellite' at the correct distance from the earth would … remain stationary above the same spot and would be within optical range of nearly half the earth's surface. Three repeater stations, 120 degrees apart in the correct orbit, could give television and microwave coverage to the entire planet."

Rockets, the letter continued, such as the V-2s that Hitler had been using to bomb England, could be used after the war to aid in the exploration of outer space. The author of the letter, and clearly the father of communications satellites, was Arthur C. Clarke, who would go on to write great science fiction works such as *2001: A Space Odyssey*.

In the year 2000, a joint Russian-European communications satellite was named after Clarke. The satellite would help provide data and video broadcasting, Internet connections, high-speed Internet access, distance-learning, and transfer of software and mobile phone signals to Eastern Europe, Siberia, North Africa, the Middle East, and India.

The Syncom series got off to a difficult start, but Clarke's vision soon got its chance. *Syncom 1* was launched from Cape Canaveral on February 14, 1963. The satellite was thought to have reached orbit but contact with it was lost soon thereafter.

Dr. Jones's Corner

I had my own encounter with Dr. Clarke after my STS-98 mission. The movie *2001: A Space Odyssey*, based on Clarke's novel, had cemented my desire as a boy to some-day join the space program. I carried onboard *Atlantis* my copy of *2001*, and Marsha Ivins and I read from the novel while in orbit. After our return, I mentioned this to a friend, who relayed the story to Dr. Clarke. Shortly after, I got a very nice note from the author, who was delighted to hear that his work had entertained astronauts on the first space shuttle mission of 2001.

Syncom 2 was put into space on July 26, 1963, and became the first satellite to be success-fully placed in a *geostationary orbit*.

The third Syncom satellite went into orbit on August 19, 1963, and it, too, was successful. *Syncom IV* was not put into space until years later, launched by the space shuttle on mission STS-41. Hundreds of geosynchronous satellites have been launched since Clarke's idea was proven practical by Syncom.

Telstar

Telstar, which was made and owned by AT&T, was launched in 1961. It was put into an elliptical orbit that allowed it to circle the earth once every two hours and 40 minutes.

In 1963, *Telstar 2* transmitted the first transatlantic color television program. That fall, following the assassination of President John F. Kennedy, satellite technology was used to broadcast his funeral around the globe.

In addition to the early communications satellites, or "comsats," already mentioned, NASA also put into space a comsat called *Relay* in 1964 and another one called *Earlybird* in 1965. *Earlybird* was put into space for a private interest known as the Communications Satellite Corporation. The two Relay satellites, each of which were put into a geostationary orbit 22,300 miles above the Earth, were also owned by a private company—in this case, RCA.

Meteorology

One of the earliest objectives of space technology was to revolutionize the fine art of predicting the weather—known in the weather biz as *meteorology*. As we know, modern weather predictions are based on photographs taken by weather satellites.

Meteorologists had long suspected that photographs from space could help predict the weather. But this opinion was not confirmed until 1947, when an unmanned U.S. rocket carried a camera into space and verified that cloud formations were clearly visible.

Space Talk

Geostationary orbit, sometimes called geosynchronous orbit, means that a satellite is orbiting the Earth at the exact speed of Earth's rotation, so that it stays above the same location all the time. Geostationary satellites make today's satellite TV possible. The dish on your roof wouldn't be much good if it could only hook up with its satellite for a few minutes each time it passed overhead.

Dr. Jones's Corner

Not all areas could receive the signals of JFK's funeral, but the worldwide grieving fostered by the global broadcast caused a "community" feeling that was perhaps unprecedented in the history of humankind. Throughout the 1960's, a "Live via satellite" caption seen on the TV screen signified an important or cutting-edge event. Now, all of us take live global television coverage for granted.

TIROS: First Weather Satellite (1960)

Meteorologists waited 13 years for the first weather satellite to go into orbit; it was called TIROS (Television InfraRed Observational Satellite). The satellite carried a video camera and made regular observations of the atmosphere below. Meteorologists found that combining their ground-based observations with pictures taken from space of an entire weather system enabled them to predict the weather with greater accuracy than ever before. In all, 10 TIROS satellites were put into orbit during the 1960s.

The *TIROS 1* satellite rocketed from Cape Canaveral on April 1, 1960, atop a Thor-Able rocket. Its mission was to test TV systems that could lead to a worldwide weather-satellite information system. It also had aboard a new system for orienting itself, based on sensing the Sun and the Earth's horizon.

Cosmic Facts

TIROS 1 was 42 inches in diameter, 19 inches high, and weighed 270 pounds. The craft was made of aluminum alloy and stainless steel. It was covered by 9,200 solar cells that charged the on-board batteries. Three pairs of solid-propellant spin rockets were mounted on the base plate. Those rockets kept the satellite spin-stabilized, like a whirling top.

Space Talk

Low resolution and **high resolution** refer to the amount of detail that can be distinguished, or "resolved," in a photograph. High-resolution photographs are more detailed and require either a low orbital altitude or a small telescope attached to the camera. A high-resolution camera zooms in on particular features, but can't cover the wide area of its "low-rez" counterparts.

TIROS 1 had two television cameras, one with *low* and one with *high resolution*. Each camera also had a magnetic tape recorder for storing photographs while the satellite was out of range of the ground station network.

The antennas on *TIROS 1* were four rods that extended from the base plate to serve as transmitters, and one vertical rod from the center of the top plate that was the receiver. Although the satellite functioned for less than three months, NASA got its money's worth out of it. Thousands of pictures of cloud formations were transmitted back to Earth, enabling meteorologists to study the relationship between the way things looked from space and the way the weather behaved on Earth. *TIROS 1* was proof that the weather could be studied and tracked from space.

TIROS 2, which went into space on November 23, 1960, atop a three-stage Delta rocket, continued the program. This satellite, which functioned for more than a year, tested new *infrared* (or heat-sensitive) equipment, which could distinguish between clouds and Earth's surface at night. It also tested another "attitude-control system" which used Earth's magnetic field to keep it oriented toward the weather below.

By the time *TIROS 9* was launched in 1965, the technology had been developed to the point that this satellite produced a "photo-mosaic" of the entire Earth, created by blending 450 photographs taken over a 24-hour period.

The TIROS program has continued right up to the present. To see the best examples of TIROS's work, watch the Weather Channel. Because of images transmitted from space, weather prognosticators today know whether it's sunny or rainy anywhere on the planet.

Photos of an Entire Hemisphere

In 1966, the U.S. put into orbit a satellite called the *ATS* (*Applications Technology Satellite*). The thing that made *ATS* unusual was that it was the first weather satellite to be placed in a geostationary orbit.

From its geostationary orbit 22,300 miles above the equator, *ATS* took the first pictures showing a whole hemisphere of the Earth at once. Using *ATS* photos, we saw for the first time how clouds moved and storms formed over wide regions.

Space Talk

Infrared light has longer wavelengths than the visible radiation that the human eye can see. Film or TV cameras that are sensitive to infrared light can show us how hot or cold things are from great distances. For example, high altitude clouds are much cooler than those near the surface, helping meteorologists understand the cloud formations they can see in orbital photographs.

Galveston

There can be no doubt about it: Weather satellites save lives. In 1900, a super-hurricane with 150-mile-per-hour winds crashed into Galveston, Texas, which sits on a low-lying barrier island on the Gulf of Mexico. The people of Galveston did not know the hurricane was coming. The National Weather Bureau did not become aware of the storm until the storm surge was already flooding the city. Tragically, more than 8,000 people died.

If that storm occurred today, the outcome would be very different, fortunately. Compare the Galveston death figure with that of a similar storm, Hurricane Andrew, which struck and devastated portions of the east coast of Florida in 1992. Although $30 billion in damage was caused by the storm, there were only 40 deaths. That's because photographs from weather satellites had given the people adequate warning and time to evacuate to safer ground.

Dr. Jones's Corner

Jamison Hawkins, chief of services at the National Weather Service's Office of Meteorology, said, "What weather satellites represented was putting a sentinel in a guard tower up above the Earth. The atmosphere is a jigsaw puzzle and without satellites, we have a lot of pieces missing."

Landsat

During the early 1970s, NASA began a program that would use the photographic techniques developed by weather satellites to study land features on the Earth. The program was called Landsat, and it was so successful that it very quickly altered the way we view our planet.

Landsat 1, manufactured by General Electric, was launched on July 23, 1972, from Vandenberg Air Force Base in California atop a Delta launch vehicle. It remained operational until 1978. By that time, it had transmitted more than 300,000 images. It demonstrated the usefulness of this technology for:

- Land surveys
- Land management
- Water resource planning
- Agricultural forecasting
- Forest management
- Sea ice movement
- Cartography (map-making)

Among the developments brought about by Landsat:

- Satellite photography could, in a few seconds, inventory crops over a large area of land. Crop forecasting for farmers, commodity traders, and insurance companies became much more precise.
- Landsat imagery of circulation and sedimentation patterns along seacoasts has been used to devise strategies for deploying equipment to contain oil spills. For example, in Japan, Landsat imagery has been used to keep track of pollution in Osaka Bay.
- Information about faults and fracture zones derived from Landsat imagery has been used in the U.S. and abroad to select locations for new power plants, and to determine the best routes for oil and gas pipelines. In Bolivia, the path of a gas pipe was changed when Landsat imagery showed an earthquake fault crossing the proposed route. A new course was determined which avoided the fault and turned out to be even shorter than the original.

Cosmic Facts

Six Landsat satellites were launched. *Landsat 6* failed. In order to compensate for it, *Landsat 4* and *5* were kept operational longer than originally intended. In fact, *Landsat 5* is still working today.

The Delta Rocket

Many of the early weather and communications satellites were launched by Delta rockets. The rocket, which is still in use today, has been responsible for the launching of more than 180 satellites. These include the TIROS and Landsat series.

The newest version of the Delta is called the *Delta II*. It's the most powerful Delta yet, standing 128 feet tall. The Delta has liquid-fueled first and second stages and a solid-propellant third stage. The average first-stage thrust, with main engine and six solid-propellant boosters, is 873,400 pounds. The U.S. Air Force and McDonnell Douglas first launched the *Delta II* from Complex 17 at Cape Canaveral Air Force Station in February 1989. (NASA relinquished control of this launch area to the Air Force in 1988.) Delta commercial launches also take place here.

The Scout Rocket

The Scout Rocket was first used in 1960 and had a long and versatile career. It was last used in 1994. During that 34-year span, more than 100 Scout rockets have been launched. They have been used for a wide variety of missions, including weather and communications satellite launches.

The final version of the Scout was a solid-propellant, four-stage booster system, 75 feet in length, with a launch weight of 46,620 pounds and a lift-off thrust of 132,240 pounds. It could boost a payload of anywhere from 400 to 465 pounds into low Earth orbit.

The lightweight Scout was launched from three different launch sites in the U.S.: Cape Canaveral, Vandenberg Air Force Base in California, and NASA's Wallops Island Flight Facility on the coast of Virginia. The U.S. and Italians also conducted joint launches from the Italian San Marco launch platform off the east coast of Kenya, Africa.

Some of the most important discoveries and innovations brought about by NASA have come in the fields of communications and weather prognostication. From its earliest days, NASA has launched communication and weather satellites, and that work continues today.

The Least You Need to Know

- Communications satellites (COMSATs) relay telephone, television, computer, and radio signals around the world.
- The first U.S. communications satellite was called *SCORE*.
- Photos of Earth from space help us monitor and predict the weather.
- Because of weather satellites, hurricane victims now have enough warning time to evacuate. The space program has saved thousands of lives and averted billions of dollars in property damage.

Part  3

Humans in Space

In the next seven chapters, you'll learn about the space explorers who blazed the trail to the Moon—the first humans in space. The Soviets put the first man in orbit and held most of the records for manned space flight, until the U.S. began to propel men out of Earth orbit toward the Moon.

You'll learn about the spaceflights of the U.S. Project Mercury, which proved that men can live and work in space. Then we'll look at Project Gemini, in which two astronauts at a time were sent into space to rehearse the skills needed for a Moon landing.

In this part, we will also look at NASA's early robotic missions beyond Earth orbit and take a break for a taste of the history and evolution of space food.

Catching Up: Yuri Gagarin, Alan Shepard, and Project Mercury

In This Chapter

- ◆ Russian man first to orbit Earth
- ◆ Alan Shepard sits atop the firecracker
- ◆ JFK targets the Moon
- ◆ Gus's sinking feeling

Not since the orbiting of *Sputnik* had the U.S. space program—and the nation—received such a kick in the pants as it did on April 12, 1961. That was the day the Soviet Union successfully put a man into orbit around the Earth and got him safely back down to the ground. The Russian success under-scored just how far behind the U.S. was in the space race.

The Soviets did not call their spacemen astronauts. They called them cosmo-nauts, "sailors of the universe." The first cosmonaut—the first human—to orbit the Earth was named Yuri Gagarin. He was launched into space from a launching pad in Tyuratam, U.S.S.R. (now Kazakhstan). Gagarin later described his liftoff into history: "The noise was approximately like the noise in an aircraft. I was prepared for much more. Then the rocket smoothly, lightly rose from its place."

"The Curvature of the Earth"

As he rose into space, Gagarin later said, "Suddenly there were no clouds. I saw the normal folds of the terrain, a region that was a little mountainous. I could see forests, rivers, ravines."

> I couldn't tell exactly where I was. I think it was the Ob, or the Irtysh, but I could see it was a large river and there were islands in it. … I could see the horizon, the stars, the sky. The sky was completely black, black. The magnitude of the stars and their brightness were a little clearer against that black background.
>
> I saw a very pretty horizon, and the curvature of the Earth. The horizon is a pretty, light blue. At the very surface of the Earth, a delicately light blue gradually darkens and changes into a violet hue that steadily changes to a black. … In my flight over the sea, its surface appeared gray, and not light blue. The surface was uneven, like sand dunes in photographs.

"No Physiological Difficulties"

"I ate and drank normally," Gagarin remembered. "I noticed no physiological difficulties. The feeling of weightlessness was somewhat unfamiliar compared with Earth conditions. Here, you feel as if you were hanging in a horizontal position in straps."

Astro Bio

Yuri Gagarin   Gagarin was born in 1934 on a collective farm west of Moscow. He originally went to trade school but his grades were good enough to get him into technical school in Saratov—the same city he looked down upon as he floated back to Earth under his parachute. He learned to fly in technical school and qualified for the Air Force. Gagarin graduated with top honors from an officer candidate school at Orenburg in 1957 and was a pilot in the Soviet air force when he was chosen to be a cosmonaut.

Following his trip into space, Gagarin understandably became the most beloved person in the Soviet Union. Sadly, Gagarin died on March 27, 1968, in a crash of a MiG-15 fighter jet during a routine test flight. In Gagarin's honor, a huge monument was erected on a main street leading to the center of Moscow. The monument depicted a man, standing erect and wearing armor, his arms spread like a diver, standing above a model of the Earth.

Gagarin orbited the Earth just once. Only 76 minutes into his mission and he was planning for re-entry! The firing of his retro-rockets—designed to slow Gagarin's spacecraft—took place under automatic control.

Tumbling Home

The firing of those rockets, although slowing Gagarin's spacecraft, also unexpectedly put it into a roll, doing a complete revolution every 12 seconds. A portion of his spacecraft which was supposed to separate before re-entry failed to do so. It wasn't until the heat of re-entry burned through a cable that the dangling section broke free from Gagarin's capsule. During re-entry, his craft stopped tumbling on its own. But there was still one more hurdle to overcome.

Bailing Out

Unlike the Americans, who would aim their capsules for an ocean recovery, the Soviets did not aim Gagarin for the water. Instead, they brought him down over the steppes of Kazakhstan. The reason was simple—much of Russia is landlocked, and the Soviets did not have a large naval recovery force. The ocean presented more problems than opportunities.

In another contrast with American astronauts, Gagarin did not land with his spacecraft, which would have been too heavy to safely land on solid ground, even with parachutes. So, at an altitude of 23,000 feet, an automatic explosive device blew the hatch off Gagarin's capsule. His ejection seat fired next, and he left his spacecraft behind, floating to the Earth with his own parachute deployed above him. To assure certification of the world record for the first manned orbital flight, the Soviets kept this ejection and parachute recovery method a secret. Gagarin landed in a field near the city of Saratov.

Two Theories of Manned Space Travel

Manned space travel was already on everyone's mind when NASA was born, but there was a lot of disagreement about how to do it. Wernher von Braun, for example, wanted to seal a man into a small nose cone, fix the cone to the top of a Redstone rocket, and blast the man to a height of 150 miles.

The Air Force, on the other hand, wanted to develop the Atlas rocket and use it to put a man in a low orbit around the Earth. Von Braun's idea was criticized as being too circus-like, like shooting a man out of a cannon—but both ideas had their day. Von Braun's concept had the advantage of being immediately feasible, whereas there were several unknowns to be tackled before the Air Force's plan could be implemented. One of NASA's first priorities when formed was to bring in all these proposals under a single program to get a man into space.

Project Mercury

NASA called the new man-in-space effort Project Mercury, after the speedy messenger of the gods in Roman mythology. The project was first announced in 1958 and soon thereafter, NASA began a search for qualified crewmembers. President Eisenhower had decreed that they be chosen from a talent pool of college-educated, military test pilots. Following an extended qualification process, the first seven U.S. spacemen—who would come to be known as astronauts (which means "sailors of the stars")—were selected in April 1959.

Designing the Mercury Capsule

Only three months after Project Mercury's conception, prototypes for the Mercury capsule were being built. The capsule was the spacecraft that the astronaut would ride in. That capsule would sit atop a rocket at launch and then separate itself from the rocket once in space.

The astronauts were involved in the planning and development of their spacecraft, and some even came up with innovations that later saved their lives. The original design, for example, lacked a window, thus making it impossible to steer the craft should the automatic guidance system malfunction.

The Mercury spacecraft, or capsule as the designers insisted on calling it, was based on a design developed by aerodynamicist Max Faget, at the National Advisory Committee for Aeronautics' Langley Aeronautical Laboratory near Norfolk, Virginia. The shape was a truncated cone with a blunt nose and broad base, unlike any other high-speed aircraft.

Faget knew that a spacecraft, like a meteorite, was apt to burn up when entering the Earth's atmosphere, so a way had to be found to both slow the capsule down and protect it from the heat. Faget protected the base of the spacecraft with a gently curving shield. The capsule was designed to re-enter the Earth's atmosphere backward, using the shield to provide drag, and carry the intense heat away from the spacecraft by "ablating," or charring away.

The Mercury 7

The Mercury astronauts were announced in April 1959, slightly more than two years before the first manned mission. They were, in the order that they eventually went into space:

- Alan Shepard
- Gus Grissom
- John Glenn
- Scott Carpenter
- Wally Schirra
- Gordon Cooper
- Deke Slayton

When the original seven astronauts were announced many thought it strange that there was one name missing: Chuck Yeager. Yeager was the nation's top test pilot and had been the first man to break the sound barrier in 1947. He didn't make the cut for a simple reason. NASA had decreed that all astronauts needed a Bachelor's degree, and Yeager had never been to college.

At the first press conference, John Glenn showed the wholesome personality and instincts for public relations that would later make him a successful politician. During the press conference, he spoke of God, motherhood, apple pie, and space travel—all in the same sentence.

NASA knew that the astronauts were the public relations key to all space programs, manned or otherwise. As it was commonly put, they "put a face on space." With the men as the focal point, it would be easier for NASA to get its budget approved by Congress. "No bucks without Buck Rogers" was the way some put it.

Accordingly, NASA was very careful about the way the seven Mercury astronauts were portrayed in the press. To protect the image of the astronauts, Lieutenant Colonel John "Shorty" Powers was named their press agent. Though the astronauts, like all men, had their flaws, the press wrote about them in glowing, hyperbolic terms. *Time* magazine called the Mercury astronauts "seven men cut from the same cloth as Columbus, Magellan and Orville and Wilbur Wright."

Cosmic Facts

Novelist Tom Wolfe turned the story of the Mercury 7 into a best-selling book, and later a hit movie, titled *The Right Stuff.* It's a popular history, though, and the astronauts themselves felt that Wolfe embellished some of the details in his tales of the Original Seven.

Monkey Business: The Flight of Ham

Two monkeys and a chimpanzee named Ham rode the Mercury capsule before the first man, to test the reliability of the spacecraft systems in action. Like the first two astronauts who would follow, Ham was blasted into space atop an 80-foot tall Redstone rocket, and the Mercury capsule parachuted back into the ocean for retrieval. Because of a malfunction, Ham's capsule re-entered Earth's atmosphere 1,400 m.p.h. faster than planned, and the chimp had to endure nearly 15 g's. When the chimp came back still breathing, NASA scientists were pretty sure that an astronaut would survive as well. Ham could at least take comfort in the fact that he fared much better than Russia's pioneering space dog, Laika.

The chimpanzee Ham has sensors attached to his body before his trip into space in 1961.

(NASA)

Freedom 7: A Cannon Shot Downrange

NASA decided that their first and second manned flights would follow von Braun's measured approach to spaceflight. The Redstone booster, not powerful enough to put a man into orbit, would instead blast him aloft on a short mission that would last only 15 minutes. Only after these first missions would NASA commit an astronaut to the tricky Atlas booster, with its orbital capability.

Only minor modifications were necessary to the Redstone, which used liquid oxygen and alcohol as propellants, to prepare it for its Project Mercury duties. The rocket would send the astronaut on a "ballistic trajectory." In other words, he would go up, arc through space like a cannonball, and then fall back to Earth.

Chosen to be first in space (on mission Mercury-Redstone 3) was Alan Shepard, the man who, years later, would stroke the most famous 7-iron shot in golf history. The Mercury astronauts were all allowed to name their own spacecraft. Shepard named his *Freedom 7*.

Alan Shepard, the first American in space.

(NASA)

Cosmic Facts

One difference between Shepard's flight and that of Yuri Gagarin was that Shepard's would take place on live television, with the whole world watching. As would become customary, all three major television networks (ABC, NBC, CBS) interrupted regular programming to show Shepard's entire mission. Not until Gordon Cooper's 22-orbit mission in 1963 would there be regular programming on network TV while an American was in space.

Cosmic Facts

The Mercury capsule and the Redstone rocket were tested separately for the most part. The rocket was erected on the launch pad long before the mission, while the capsule was lifted atop the rocket and secured in place on the launch pad for the final set of tests only days before the lift-off.

Before his flight, Shepard had this to say:

> I think the one thing that strikes me as I look back on [my two-year] training program is that I have really developed a feeling of confidence—a confidence in the people with whom I work, and confidence in the systems with which I am dealing and will have to deal with in flight, and, of course, a confidence in myself.

Before Shepard blasted off, the electrical, mechanical, and communications systems in his spacecraft had been tested and double-tested—as had the life-support and pressurization systems connected to Shepard's space suit. A complete simulated launch—a dress rehearsal, if you will—had been run to make sure that all the systems worked in conjunction with one another. As Shepard climbed into his capsule, he knew that everything was as ready as it could possibly be. Yet Shepard, like the other astronauts, was also keenly aware that both his booster and capsule had been built by the lowest bidder!

In case there was a problem during the early stages of powered flight—many rockets had blown up just after blast-off—the Mercury capsule was topped with a small emergency rocket mounted atop a small tower. If a problem occurred, the astronaut could press a button, ignite the escape rocket, and the capsule would be pulled clear of the launch vehicle for a landing via parachute.

Shepard's flight was originally scheduled to begin on May 2, 1961. But, hours before lift-off, the mission was postponed because of poor weather. The flight was rescheduled for 72 hours later.

Dr. Jones's Corner

The structure that stands next to a rocket, both to support it and allow technicians to work on it, is called a gantry. The gantry contains an elevator for getting to the top of the rocket. (On my shuttle missions, I rode the same elevator that Neil Armstrong rode as he set off for the moon.) The small room on the gantry that provides access to the spacecraft hatch is called the "White Room," after the color of the original Mercury facility. The gantry for the Mercury missions was mounted on rails and was rolled away from the rocket a few hours before the launch.

The Redstone rocket boosted the first two Americans— Alan Shepard and Gus Grissom—into space.

Mercury-Redstone 3 would only blast off following a 10-hour countdown. The countdown was divided into two parts. Four hours were completed on the day before the blast-off, with the final six leading right up to the lift-off the next day. This was done to reduce fatigue and the chance of mistakes, as the countdown's purpose was to check and double-check all of the booster and spacecraft systems.

"What a Beautiful View!"

Shepard's flight finally came on Friday, May 5, 1961. The first American in space traveled 116 statute miles high, and 302 miles east of Cape Canaveral; *splashdown* was in the Atlantic Ocean. During his short trip, one of Shepard's duties was to test the manual *pitch* (ability to move up and down), *yaw* (side to side), and *roll* controls while in space, to see how well he

Space Talk

For either an emergency or normal landing, naval forces had the job of retrieving the capsule and astronaut. Helicopters would pluck both astronaut and capsule out of the sea following his wet landing. The landing of the astronaut, still in his capsule, in the ocean came to be known as **splashdown.** My astronaut colleagues and I still celebrate the end of each successful mission with a "splashdown" party in Houston.

could accomplish simple maneuvers under weightless or free-fall conditions. Shepard, the expert test pilot, had no trouble executing the maneuvers.

During his brief moments in space, Shepard confirmed that the cloud patterns beneath him resembled the weather maps he had previously been shown: "On the periscope," he said. "What a beautiful view! Cloud cover over Florida is three to four tenths near the eastern coast—obscures up to Hatteras. ... I just saw Andros Island, identified the reefs."

As he stole looks at the Earth, Shepard's blood pressure, heart rate, respiration rate, and body temperature were monitored carefully by telemetry. (Telemetry is the science of measuring a quantity, and then transmitting that data to a distant location.) The entire flight took 16 minutes and offered Shepard about five minutes of weightlessness.

Astro Bio

Alan Shepard Alan B. Shepard Jr., was born on November 18, 1923, in East Derry, New Hampshire. He attended the U.S. Naval Academy, where he received his Bachelor of Science degree in 1944. After graduation, he started his naval career in the Pacific during World War II aboard the destroyer *Cogswell*. After the war, Shepard learned to fly and received his wings in 1947. He was assigned to Fighter Squadron 42 with which he served several tours aboard aircraft carriers in the Mediterranean. In 1950, he attended U.S. Navy Test Pilot School in Maryland. After graduation, he became a test pilot, logging more than 8,000 hours flying time by the time he became one of the Mercury 7.

Following his suborbital flight—the first U.S. manned space mission—Shepard was named Chief of the Astronaut Office, where he was responsible for coordinating the development and implementation of astronaut training programs. An inner-ear disorder took him off flight duty for a time, but in May of 1969, after a corrective ear operation, he was returned to full flight status. Shepard's second flight had been a long time coming, but it was worth the wait: He walked on the Moon as commander of *Apollo 14*, aloft from January 31 until February 9, 1971. Shepard retired from NASA in 1974 and went into private business in Houston, Texas. He died of cancer on July 21, 1998.

President Kennedy's Promise

Just 20 days after Alan Shepard's flight, President John F. Kennedy addressed a joint session of the U.S. Congress to discuss "urgent national needs" in the face of recent Communist advances. Here is what JFK said:

> If we are to win the battle that is going on around the world between freedom and tyranny, if we are to win the battle for men's minds, the dramatic achievements in space which occurred in recent weeks should have made clear to us all, as did the *Sputnik* in 1957, the impact of this adventure on the minds of men everywhere who are attempting to make a determination of which road they should take. Now it is time to take longer strides, time for a great new American enterprise, time for this nation to take a clearly leading role in space achievement, which in many ways may hold the key to our future on Earth.

> *I believe that this nation should commit itself to achieving the goal, before the decade is out, of landing a man on the Moon and returning him safely to the Earth.* No single space project in this period will be more exciting, or more impressive to mankind, or more important for the long-range exploration of space; and none will be so difficult or expensive to accomplish.

Cosmic Facts

Only 11 minutes after splashdown both Shepard and *Freedom 7* were aboard the rescue vehicle, the carrier *Lake Champlain*. Moments later Shepard received a congratulatory phone call from President John F. Kennedy.

It was President John F. Kennedy who said, "I believe that this nation should commit itself to achieving the goal, before the decade is out, of landing a man on the Moon and returning him safely to the Earth."

(NASA)

At the time Kennedy made this commitment to go to the Moon, the United States had a grand total of 16 minutes of human spaceflight experience.

Liberty Bell 7

Virgil "Gus" Grissom became the second American in space on July 21, 1961, in a Mercury capsule he called *Liberty Bell* 7, after the obvious shape of his spacecraft. He even had a crack, like that on the famous bell, painted onto its exterior. Grissom, like Shepard before him, was launched by a Redstone rocket, and his suborbital flight would propel him into space, but not into orbit.

New and Improved Capsule

Grissom's capsule was new and improved. It had a redesigned control panel, now capable of accommodating future orbital flights. The window was much larger so Grissom would have a better view, and there was an escape hatch that could be jettisoned using explosive bolts.

Dr. Jones's Corner

Let's talk about that unique sensation that Gus and Alan experienced in space—weightlessness or zero-gravity. Zero-gravity may be what it feels like, but the reality is that gravity is present—it's what's keeping your spacecraft in orbit. In fact, gravity is only a little weaker two hundred miles up in space than it is at Earth's surface. I think of our condition as "free-fall," the absence of perceived external forces experienced by falling bodies (good old Newton's Laws again). As we fall (in orbit, or in a sub-orbital toss), we seem to feel no external forces—we simply float. Now, it's not entirely an absence of force, or acceleration; it's just that these accelerations are tiny compared to what we are used to living on the surface of our planet. Nearly 100 percent of the time, astronauts are living under conditions loosely termed "microgravity."

Whereas Shepard had had no choice but to stay inside his spacecraft until he was pulled out by the recovery team, Grissom wanted to have the option of leaving the capsule on his own, in case something was wrong—on fire or sinking, for example. The hatch was secured by 70 explosive bolts and could be activated by the astronaut by pressing a plunger in the cockpit.

Cosmic Facts _____

With the space race in full swing, and NASA's budget larger than ever, there was great growth in the organization—and not just in Cape Canaveral. Four new facilities were built. NASA purchased 125 square miles of land on Merritt Island, just north of the Cape, and there built a 130-million-cubic-foot "Vehicle Assembly Building," where the rockets that would take man to the Moon would eventually be built. The Manned Spacecraft Center was built 20 miles outside Houston, Texas, in a land deal involving Rice University, Humble Oil, and political friends of Vice President Johnson.

There had been two holds in the countdown on the morning of Grissom's launch, one for repair of that hatch, and another when cloud cover blocked the view of cameras set up to record the flight.

Once aloft, Grissom's flight went smoothly, although he did complain that his spacecraft's attitude control felt "sticky and sluggish."

Astro Bio

Virgil Ivan "Gus" Grissom   Gus Grissom was born April 3, 1926, in Mitchell, Indiana. The son of a railroad worker, he was the oldest of four children. In school he excelled at math but was run-of-the-mill in his other subjects. He enlisted as a high school senior as an aviation cadet and began his training immediately following graduation. Japan surrendered before he could learn to fly, and he found himself working a desk job in the Army Air Corps. He was discharged in November 1945. Soon thereafter he enrolled at Purdue University and earned his bachelor's degree in mechanical engineering in three and a half years. By this time Gus's heart was set on becoming a test pilot, so he reenlisted in the Air Force. He earned his wings in time to be shipped out to Korea, where he completed 100 combat missions.

Following the Korean War he became a flight instructor. He then attended test pilot school at Edwards Air Force Base and received his test-pilot credentials in 1957. He hadn't been a test pilot for long when he volunteered for the Mercury astronaut competition. Following the battery of medical and aptitude tests, he was chosen to be one of the Original Seven.

Gus's waterlogged experiences on his Mercury flight obviously didn't hurt his reputation inside NASA. Grissom became the first American to go into space twice—as commander of _Gemini 3_ on March 23, 1965. Tragically, he was killed in the _Apollo I_ fire in 1967 (see Chapter 18, "Early Apollo Missions"). At that time he was a leading favorite to be chosen to command the first manned landing on the Moon.

How Was *Liberty Bell 7* Unlike Ivory Soap?

It was only after a successful splashdown in the Atlantic that things began to go wrong. According to Grissom's own statement:

> I opened up the faceplate on my helmet, disconnected the oxygen hose from the helmet, unfastened the helmet from my suit, released the chest strap, the lap belt, the shoulder harness, knee straps and medical sensors. And I rolled up the neck dam of my suit … As soon as I had finished looking things over, I told Hunt Club (the recovery helicopter) that I was ready. According to the plan, the pilot was to inform me as soon as he had lifted me up a bit so that the capsule would not ship water when the hatch blew. Then I would remove my helmet, blow the hatch and get out. … Then suddenly, the hatch blew off with a dull thud.

Ocean water rushed into the capsule and Grissom—who could not explain why the hatch had blown—crawled out so as not to go down with the ship. While the helicopter was able to rescue him before his suit filled with seawater and dragged him under, *Liberty Bell 7* went to a watery grave. It was only recovered from the seafloor, 16,000 feet down, in the summer of 1999.

Talking later about the sinking of *Liberty Bell 7*, Grissom said: "It was especially hard for me, as a professional pilot. In all of my years of flying—including combat in Korea—this was the first time that my aircraft and I had not come back together. In my entire career as a pilot, *Liberty Bell* was the first thing I had ever lost." Grissom's fellow astronauts always maintained that the hatch had blown accidentally. Tragically, Grissom's fear of a recurrence of the accident contributed to his death in the *Apollo I* fire, as that craft lacked an explosive hatch that could be blown from the inside.

The Least You Need to Know

- Yuri Gagarin—a Russian—was the first man in space. He orbited the Earth in April 1961.
- Alan Shepard was the first American in space (May 1961).
- The first two American astronauts—Shepard and Grissom—were launched atop Redstone rockets on short ballistic hops.
- Following splashdown, Gus Grissom's spacecraft filled with water and sank when the escape hatch blew. Grissom himself nearly drowned.

Around the World in 90 Minutes: Astronauts in Orbit

In This Chapter

- John Glenn goes into orbit
- *Friendship 7's* rocky return
- A hero's welcome
- Tough act to follow

An American astronaut did not complete an orbit around the Earth until NASA's third piloted mission. The star of the show was the same man who had been a scene-stealer at the Mercury 7's first press conference with his mom and apple pie talk about space exploration: John Glenn.

Friendship 7: First U.S. Orbital Flight

Friendship 7, sitting atop Mercury-Atlas 6, endured 11 delays in its countdown. Malfunctions, improvements, and bad weather made Glenn's long wait to fly seem endless. He finally made it into space on February 20, 1962.

John Glenn onboard
Friendship 7.

(NASA)

The first part of the mission went smoothly. The Atlas rocket performed perfectly and Glenn's *Friendship 7* was good for at least six orbits. He was treated to a spectacular view of the Earth, one that no American had ever seen before.

"Fireflies": The Luminous Particles

Glenn started a small mystery when, just as he was viewing his first of three sunrises, he radioed down that he saw fireflies outside his window. He said they looked like luminous fireflies dancing all around his capsule. Some thought at first that Glenn had discovered life in space, others that he was hallucinating, but the best theory now is that he saw small flakes of ice that had been dislodged from the surface of his spacecraft by the sunshine.

But it wasn't long before there were more serious concerns. The automatic attitude control system began to malfunction near the end of Glenn's first orbit.

A Brick and a Stick

Glenn, who had argued long and hard for a manual control system in the Mercury spacecraft—a "brick and a stick" they called it—in case the automatic system failed, had saved his own life with his foresight. He used the manual piloting system for his remaining two orbits of the Earth, and for his re-entry.

Cosmic Facts

Prior to his own flight in *Friendship 7*, Glenn had been the backup pilot to both Alan Shepard and Gus Grissom.

Astro Bio

John Glenn   John Herschel Glenn Jr., was born on July 18, 1921, in Cambridge, Ohio, and grew up in New Concord, Ohio. He attended Muskingum College, where he received a B.S. in engineering. After school, Glenn was accepted into the Naval Aviation Cadet Program. He won his wings and lieutenant's bars in 1943 as a Marine. Glenn flew 59 combat missions in the South Pacific.

After the war, Glenn trained other pilots and did some test-pilot work. During the Korean War, Glenn flew 63 ground-support missions. He was credited with shooting down three North Korean MiGs in nine days. When the conflict ended, Glenn was accepted at the Navy Test Pilot School. He tested Navy and Marine Corps jet fighters from November 1956 to April 1959. In July 1957 he piloted the first transcontinental flight to average faster than the speed of sound, setting a transcontinental speed record (3 hours, 23 minutes, and 8.4 seconds).

While working as a test pilot, he was asked to help design what would become the Mercury capsule. He also agreed to test the centrifuge and other equipment that NASA would use to simulate the g-forces an astronaut would have to endure to re-enter the Earth's atmosphere. As a candidate for the Mercury 7, Glenn had already tested and contributed to the design of the equipment he would be tested with during the selection process.

Less than two years after his historic space trek, Glenn realized that his age (he was in his early forties) was apt to keep him from ever walking on the Moon. (Many in NASA also believe that President Kennedy personally ordered that Glenn not be assigned another flight, because of the risk of losing a national hero.) Glenn resigned from the Manned Spacecraft Center in January 1964 and announced his plans to run for the Senate in Ohio. He dropped out of the senatorial race after a fall that left him with a concussion and a lingering inner ear problem. He again ran for the Senate in 1970 but was defeated. He tried again in 1974 and the third time was the charm.

In 1998, Glenn became the oldest man to orbit the Earth (see Chapter 26, "Return to Flight").

But it wasn't clear sailing yet. As the time approached for Glenn to re-enter the Earth's atmosphere, Mission Control saw indications that *Friendship 7*'s heat shield was loose.

The Burning Retropack

Controllers suspected the signal was false, but they couldn't discount the risk. If the heat shield tore loose during re-entry, both the capsule and Glenn would be incinerated.

To assure that the heat shield remained in place during re-entry, the *retropack* (the assembly containing the rockets that were fired to slow the capsule, normally jettisoned after firing) was left on, in hopes that its straps would help hold the heat shield in place.

As the capsule was enveloped in the searing fireball of re-entry, Glenn could see portions of the burning retropack flying past his window. As he concentrated on steering the capsule home, the heat shield held.

Nation Holds Its Breath

Glenn, reduced to using the "brick and a stick" he had fought for, while checking his course through the window engineers hadn't wanted to give him, knew that his angle of re-entry was crucial. If he came in too steeply, he would burn up, as the temperatures would be too great for even an intact heat shield. If he came in at an angle that was too shallow, he would skip off the Earth's atmosphere and head off into a skewed orbit without a way to maneuver back for re-entry.

Glenn had to thread a needle of life and death—and he had to do it at almost 18,000 miles per hour.

Americans who are old enough to remember that day recall the horrible moments when, as the re-entry fireball around Glenn was at its hottest, all communications were lost with the spacecraft.

We did not yet know that this was a normal part of re-entry, as the hot plasma surrounding the capsule blocked radio transmissions. Until he emerged from the fireball, there was no way to know whether he and the capsule had burned up. But *Friendship* 7 survived, and by the time Glenn's parachute opened, he regained radio contact. After a wait that seemed to take forever, the nation could breathe easily again.

Glenn landed in the Atlantic approximately 800 miles southeast of Cape Canaveral near Grand Turk Island, 41 miles west and 19 miles north of the planned landing target.

The astronaut and his spacecraft were picked up by the destroyer *Noa*. Glenn and *Friendship* 7 were aboard the ship only 21 minutes after splashdown.

Record Amount of Tickertape

John Glenn received an unprecedented hero's welcome after his return from space. He didn't have to go to Washington to visit President Kennedy. Kennedy came to him. When Glenn returned to Cape Canaveral, JFK was there to greet him—and present him with the Space Congressional Medal of Honor.

Then Glenn and his family headed north, first to Washington, D.C., where Glenn gave a speech to a joint session of Congress. Then it was on to Broadway in New York City, where he was given a tickertape parade. Onlookers threw a record amount of paper (3,500 tons) as Glenn passed, breaking the previous record (3,249 tons) held by the first man to fly across the Atlantic Ocean solo, Charles Lindbergh.

Glenn's capsule, *Friendship* 7, went directly to the Air and Space Museum at the Smithsonian Institution in Washington, D.C., where it was displayed in the main hall along with Charles Lindbergh's plane, *The Spirit of St. Louis*. Today, in the "Milestones of Flight" gallery, Glenn's capsule sits just beneath the first airplane, the Wright Flyer.

Cosmic Facts
Friendship 7 Facts
Launch: February 20, 1962
Length of Flight: 4 hours, 55 minutes, and 23 seconds
Apogee: 162 statute miles
Perigee: 100 statute miles
Orbital velocity: 17,500 miles per hour
Total distance traveled: 75,679 statute miles
Maximum "g's": 7.7

Dr. Jones's Corner
After receiving his medal, Glenn took time to show his wife and children his capsule, and how he had threaded the needle with it as he re-entered the Earth's atmosphere without burning up.

Aurora 7: Carpenter's Flawed Mission

Scott Carpenter was the next American to orbit Earth. He had a hard act to follow, and since his mission was basically a repeat of Glenn's, people tend to forget about it. It didn't help that his mission seemed slightly jinxed.

The purpose of his mission was to determine if a man could work in space. Like John Glenn before him, Carpenter was slated to go around the Earth three times. During his flight, he was to conduct numerous experiments, such as deploying a tethered balloon and observing flares fired on Earth.

Carpenter blasted off atop an Atlas rocket on May 24, 1962, sitting in his spacecraft, *Aurora* 7. His *apogee* was 164 miles. While orbiting Earth, he traveled at 17,532 miles per hour.

Space Talk

The **apogee** of an orbit is its maximum altitude above the Earth. (The perigee is an orbit's closest approach to Earth.)

Dr. Jones's Corner

One successful experiment accomplished during the mission was the ranking of colors by their visibility in space. Carpenter judged orange to be the clearcut winner (maybe that's how Tang got picked).

The Bubble Bursts

The balloon deployment did not go well. The experiment was designed to measure the drag on the balloon in the thin atmosphere and observe its behavior. It was painted in various colors and Carpenter was to determine which were most easily visible in space.

The balloon failed to inflate properly, and never achieved its maximum size. The timing of the experiment was thrown off when it took longer than expected to reach the end of its 100-foot nylon tether.

Next, the balloon jettison didn't work properly, and *Aurora* 7 continued to drag the balloon along behind. It burned up upon re-entering the atmosphere.

Pilot Errors

Carpenter also made several mistakes during his three orbits that affected the success of the mission. He accidentally spent too much fuel when, switching from his manual to fly-by-wire system, he neglected to turn off the fuel flow. When it came time to re-enter the Earth's atmosphere, he made an error in positioning the capsule, an error in yaw, so that his capsule was canted 25 degrees to the right of where it should have been. Because of the error in positioning, the capsule did not slow down as much as it should have when Carpenter fired his retro-rockets. That error was made worse when Carpenter, behind schedule in his checklist, fired the retros three seconds late.

Cosmic Facts

Folks who are prone to motion sickness generally don't become astronauts in the first place, but there's no way to anticipate the effects of prolonged near-weightlessness on an individual. The so-called "Vomit Comet"—a weightless training aircraft—can help you get familiar with free-fall for a short period of time, but learning to live under weightless conditions can only be practiced by doing it for real. Research tells us that about 35 percent of astronauts experience a motion-sickness-like ailment known as Space Adaptation Syndrome. If untreated, it goes away on its own after two or three days of queasiness. Today, fortunately, we have an effective anti-nausea drug that brings quick relief.

Astro Bio

Scott Carpenter   Born on May 1, 1925, in Boulder, Colorado, to a doctor and his wife, Malcolm Scott Carpenter was raised by a family friend after his parents separated and his mother became ill. Following high school, Carpenter spent a year at the Navy's V-5 flight training program at the University of Colorado. He then trained for six months at St. Mary's Pre-flight School in Moraga, California, followed by four months in primary flight training at Ottumwa, Iowa. He received his degree in aeronautical engineering from the University of Colorado in 1949. Carpenter then joined the Navy and received flight training from November 1949 to April 1951 at Pensacola, Florida, and Corpus Christi, Texas.

During the Korean War, Carpenter engaged in anti-submarine patrol, shipping surveillance and aerial mining activities in the Yellow Sea, South China Sea, and the Formosa Straits. He entered the Navy Test Pilot School at the Naval Air Test Center, Patuxent River, Maryland, in 1954, and, after completion of his training, was assigned to conduct flight test projects with the A3D, F11F, and F9F and assisted in other flight test programs.

When NASA was formed, Carpenter was among the 110 military pilots chosen as candidates for America's first manned venture into space. On April 9, 1959, he was chosen as one of the seven Mercury astronauts. He served as John Glenn's backup pilot for America's first manned orbital flight. When NASA grounded Deke Slayton because of his heart condition, Carpenter was selected as prime pilot for the next mission.

Tense Re-Entry Redux

The unexpected angle of re-entry, compounded by the dangerously low level of fuel, meant that, as Carpenter put it, the window to re-enter the atmosphere without burning up was going to be "pretty tight." *Aurora* 7 ended up overshooting its landing target by 250 miles. Keeping the horizon in view through the capsule window, Carpenter gingerly guided the capsule through its long plunge back to Earth, being careful to use as little fuel as possible. Because the capsule was misaligned during re-entry, the heat shield couldn't properly do its job. The capsule began to heat up so Carpenter put the spacecraft into a slow roll, like a rotisserie, so that the craft would be heated equally on all sides.

Looking out the window, Carpenter could see an orange ring of fiery particles stretching out behind his vehicle. Tiny pieces of the heat shield were melting way and carrying heat away with them, causing the light show.

When the capsule began to shake violently, Carpenter used the last of his fuel to control it. For a time he feared that he would re-enter nose down, an attitude that would have destroyed his parachute upon deployment.

Cosmic Facts _____

Bobbing next to his capsule in a small raft, Scott Carpenter was rescued from the sea only after he signaled the search planes by using his survival mirror, reflecting sunlight toward patrolling aircraft.

Because of his concerns, Carpenter deployed the small drogue parachute at 26,000 feet rather than at 21,000 as scheduled. The main canopy failed to deploy automatically, too, and had to be activated manually by the astronaut.

The main chute then deployed perfectly and—after 4 hours, 53 minutes, and 47 seconds of flight—*Aurora 7* drifted down into the sea, 1,000 miles southeast of Cape Canaveral, 250 miles from his nearest rescuer. He had to wait for 45 minutes before he was found, and another two hours after that before he was pulled from the sea by a helicopter from the USS *Intrepid*. The country, worried for hours that Carpenter was lost, was so relieved by his recovery that the press overlooked most of the difficulties he'd experienced during the flight.

The Least You Need to Know

◆ John Glenn was the first American to orbit the Earth.

◆ When Glenn's guidance system broke down, he coolly piloted his craft through a harrowing re-entry back to Earth.

◆ Glenn proved the value of having a pilot at the controls of his spacecraft.

◆ Scott Carpenter followed Glenn into space on a flawed but successful mission, proving that the U.S. was getting the hang of orbital flight.

Near Perfection

In This Chapter

◆ Wally Schirra's six-pack of orbits

◆ A lengthier stay in space

◆ Gordon Cooper takes a nap

◆ Twenty-two orbits

It was as if all the bugs in Project Mercury had worked themselves out during Scott Carpenter's bad-luck mission. There were two Mercury flights left, and they would be close to flawless.

Sigma 7: Engineering Precision

NASA announced on June 27, 1962, that Wally Schirra had been designated for America's fifth manned space mission, Mercury-Atlas 8. His launch was originally scheduled for September 28, 1962. That day came and went, however; a bad fuel-control valve caused a delay until October 3.

Schirra had named his spacecraft *Sigma* 7—the Greek letter Sigma (Σ) designating "engineering precision." This turned out to be a fitting symbol for this flight, when both spacecraft and astronaut turned in a near-perfect performance.

Cosmic Facts

Schirra's six-orbit flight—during which he achieved an altitude of 175 miles and a speed of 17,557 miles per hour—lasted 9 hours, 13 minutes, and 11 seconds.

Dr. Jones's Corner

Wally Schirra's father had an engineering degree from Columbia University and had flown bombing missions over Germany during World War I. After the war, Wally's mother and father barnstormed together as a stunt flier act. Mom stood on the wing of the biplane while Dad handled the flying.

Schirra was the consummate test pilot at work, ticking off the steps on his flight plan with precision. He accomplished the mission's premier goal when he proved that he could successfully manage electricity and fuel during a flight—a necessary skill for longer missions such as those to the Moon.

The precision of the flight carried right through to splashdown in the Pacific Ocean, 275 miles northeast of Midway Island. The splashdown occurred only four and a half miles from the recovery ship, the aircraft carrier *Kearsarge*.

In his flight report, Schirra wrote:

> One of the main objectives of this flight was an engineering evaluation of the spacecraft systems to determine their capabilities for an extended mission. In line with this objective, we wanted to demonstrate that the consumable supplies could be conserved sufficiently to permit longer duration flight in the future using the Mercury spacecraft. Of course, most of the consumables, such as water, electrical power, and contaminant filters, will have to be increased, but it is still important to determine the long-term consumption rates.

Schirra also wrote about the "camaraderie of everyone concerned with the flight preparations and equipment":

> It was certainly a thrill while entering the spacecraft on launch day to see a dummy "ignition key" on the control stick safety pin. This and other small gestures really helped to make me realize that there are many other people who were interested in what I was doing. We know this inherently, but these visible examples of it mean quite a bit. Here again, sigma symbolizes the summation of the great efforts exerted by each and every man in the vast Mercury team. ... This spacecraft, the crew that prepared her, and the flight itself, truly combine to make this MA-8 experience the high point in my life.

Astro Bio

Walter Marty Schirra   Wally Schirra was born in Hackensack, New Jersey, on March 12, 1923. Schirra studied aeronautical engineering at the Newark College of Engineering from 1940 to 1942. In 1942, he was appointed to the United States Naval Academy and received a Bachelor of Science degree on June 6, 1945. Upon graduation he was commissioned in the Navy as an ensign and assigned to the armored battle cruiser *Alaska*.

In 1946 he was assigned to the staff of the 7th Fleet, based off China, and in 1948, after completing pilot training at Pensacola Naval Air Station, Florida, he was assigned to Fighter Squadron 71. He flew as an exchange pilot with the 154th Fighter Bomber Squadron during the Korean War, chalking up 90 combat missions. Schirra was credited with downing at least one MiG fighter.

Schirra served as a test pilot at the Naval Ordnance Training Station at China Lake, California, from 1952–54, working on the development of the Sidewinder air-to-air missile. From 1954 to 1956, Schirra was a project pilot and instructor pilot. He attended the Naval Air Safety Officer School at the University of Southern California in 1957, and completed test pilot's training at the Naval Air Test Center at Patuxent River, Maryland, in 1959. It was there that he first heard about the Mercury Project. He applied for the job, but hit a stumbling block. Doctors discovered during tests that he had a polyp on his larynx. When they arranged in a matter of days to give him three months' worth of treatments, including the surgical removal of the polyp, Schirra knew that he was being considered to be one of the Final 7.

After making the cut, his special contribution to the team was the development of environmental controls, or life-support systems, that would ensure the safety and comfort of the astronaut within the spacecraft.

Russians Still Winning the Space Race

The success of Project Mercury to this point had done much to close the gap between the space programs of the Soviet Union and the U.S., but the Russians were still ahead—by a lot. By the time Wally Schirra finished his six revolutions around the planet, Soviet cosmonauts had already had a man in space for more than a full day and had already had two manned spacecraft in space at the same time, even orbiting within three miles of one another. The first woman had flown in space: Valentina Tereshkova, a Russian.

But Project Mercury had hit its stride, and there would be much catching up on the next mission, the most ambitious one yet.

Faith 7: A Lengthier Stay in Space

Gordon Cooper's mission was to evaluate the effects of an extended stay in space. Indeed, he was scheduled to stay in space longer than all the other Mercury astronauts combined.

Cosmic Facts

Here's a little lesson I call the "Olfactory factor," or, "The Nose Knows." Another common effect of near-weightless conditions on an astronaut's body is a change in the upper respiratory system. Bodily fluids which usually drain toward the lower torso or legs stay where they are, or even drift up toward the head. This "fluid shift" phenomenon gives some astronauts a "stuffy" feeling in their sinuses and nasal passages and causes a ruddy complexion and perhaps even a headache. This malady was not considered so bad in the early days of NASA's manned spaceflight program—it kept astronauts from having to taste the space food!

Space Talk

We say that we experience 4 **g's** during liftoff and zero g's once in space. "One g" represents the normal acceleration—or force—of gravity on us while walking on Earth. Two g's would be twice that amount of acceleration, four g's would be four times the acceleration, and so on.

Cosmic Facts

One item that came out of NASA technology is the "smart bed," which is made out of stuff called visco-elastic foam. It is a mattress with a memory. It automatically senses your body weight and temperature, and then it responds by molding to your body's exact shape and position.

Cooper, piloting Mercury-Atlas 9, left Earth on May 15, 1963. He called his spacecraft *Faith* 7. Cooper ended up circling the Earth 22 times. He was in space for 34 hours, 19 minutes, and 49 seconds.

Cooper traveled 546,167 miles at 17,547 miles per hour. The highest altitude he achieved was 165.9 miles. He endured a maximum of 7.6 *g's.*

Cooper became the first astronaut to sleep in space. With a reputation for being a relaxed fellow who could nap at will, Cooper could snooze anywhere—and he proved it by becoming the first American to catch a few ZZZZZZs in orbit.

Most astronauts have little difficulty sleeping while in space. A few need to take an occasional sleeping pill, and big mission events coming up the next day can make one restless. Also, the sensation of one's legs floating "up" off the mattress can be disconcerting at first.

One difference between sleeping in space and sleeping on Earth is that there's no sensation of weight between your body and the bed.

Astro Bio

Gordon Cooper   Born on March 6, 1927, Leroy Gordon Cooper Jr. was the son of an Air Force colonel. He graduated from high school in Murray, Kentucky, and enlisted in the Marine Corps. He served as a guard in Washington, D.C., and was on the Presidential Honor Guard when he was discharged. Cooper then attended the University of Hawaii, where he met Trudy B. Olsen, a pilot whom he later married. While in college, Cooper's ROTC experience earned him a commission in the Army.

From there he was transferred to the Air Force and, in 1949, reported for flight training at Perrin Air Force Base, Texas, and Williams Air Force Base, Arizona. After receiving his wings, Cooper flew with the 86th Fighter Bomber Squadron in Landstuhl, West Germany. After four years he became a flight commander in the 525th Fighter Bomber Squadron.

Cooper moved to Wright-Patterson Air Force Base in Dayton, Ohio, in 1954, to attend the Air Force Institute of Technology, and there received his bachelor's degree in aeronautical engineering. In 1956–57 he attended the Experimental Flight Test School at Edwards Air Force Base. While a test pilot there he was called to Washington as a Mercury astronaut candidate. He is credited with developing an emergency survival knife that was carried by astronauts, as well as systems to protect astronauts in case of accidents on the launching pad. He worked as a capsule communicator (CAPCOM) on both John Glenn and Scott Carpenter's flights. Cooper was also the commander of *Gemini V*, which spent close to eight days in space. Together, his two space flights gave Cooper 225¼ hours in space.

After a few years of working on the project that was to become *Skylab* (see Chapter 23, "Skylab and Space Détente,"), Cooper retired from both NASA and the Air Force in 1970 and formed an aviation and aerospace consulting firm based in Hialeah, Florida. Since 1980, he has been president of X-L Incorporated, a firm that develops alcohol-based aviation fuel.

In space we float, which means that we touch all sides of our own sleeping bags with the same force. It can be quite lulling, but it takes a while for some to get used to it. I find it pleasant—rather like sinking back into a very deep feather mattress. Some astronauts, though, have such a hard time settling in that they take sleeping pills to relieve their insomnia until they adapt more fully to free-fall.

Then there are astronauts like Gordon Cooper who have a knack for drifting off whenever they want and can fall asleep in space the first time they ever try. As for me, I've always been so tired at the end of the day that I'm asleep soon after my head floats off the pillow.

Dr. Jones's Corner

On some space shuttle missions half of the crew slept while the other half worked. I've flown two of these "dual-shift" missions. Astronauts slept inside a shoebox-shaped compartment. This "sleep station" was quite snug, wide enough for the shoulders, with perhaps a foot of space above the sleeper's face. This compartment bothered some more than others, but astronauts are screened for claustrophobia, and I enjoyed the privacy. The sleep station kept out light and noise, and included a sleeping bag clipped to the flat bottom of the box. (I used to sleep with my face down, my sleeping bag attached to the ceiling.) Today, crews on extended missions work a one-shift schedule—the entire crew sleeps at the same time. This keeps the crew working closely as a team; Mission Control watches over the spacecraft while the crew gets its rest.

Are There UFOs in Space?

Gordon Cooper was probably the first astronaut to have his name linked to UFOs. Twelve years earlier, while flying an F-86 Sabre over West Germany, he had seen several metallic-looking discs which flew at great altitude, and appeared to be able to outmaneuver Cooper's jet fighter.

Cooper later told the U.N.: "I believe that these extraterrestrial vehicles and their crews are visiting this planet from other planets. Most astronauts are reluctant to discuss UFOs. I did have occasion in 1951 to have two days of observation of many flights of them of different sizes, flying in fighter formation, generally from east to west over Europe."

In another interview, Cooper added:

> For many years I have lived with a secret, in a secrecy imposed on all specialists in astronautics. I can now reveal that every day, in the USA, our radar instruments capture objects of form and composition unknown to us. And there are thousands of witness reports and a quantity of documents to prove this, but nobody wants to make them public. Why? Because authority is afraid that people may think of God knows-what kind of horrible invaders. So the password still is: We have to avoid panic by all means. I was furthermore a witness to an extraordinary phenomenon, here on this planet Earth. It happened a few months ago in Florida. There I saw with my own eyes a defined area of ground being consumed by flames, with four indentations left by a flying object that had descended in the middle of a field. Beings had left the craft (there were other traces to prove this). They seemed to have studied topography, they had collected soil samples and, eventually, they returned to where they had come from, disappearing at enormous speed. I happen to know that authority did just about everything to keep this incident from the press and TV, in fear of a panicky reaction from the public.

The next astronauts to see something unexplainable in space were astronauts Jim McDivitt and Ed White in June 1965. This was the mission in which White made the first U.S. space walk. Their Gemini spacecraft was traveling over Hawaii when they saw a metallic object. The object had arms sticking out of it. McDivitt reportedly took pictures of the object but those photos have never been released.

Only a few months later, in December 1965, astronauts James Lovell and Frank Borman saw a UFO during their second orbit around the Earth. Borman reported seeing the unidentified spacecraft and was told by ground control that he was probably seeing his own Titan booster rocket which had been jettisoned. Borman, however, said that he could see both the UFO *and* the Titan booster. Later in the flight, Lovell too reported seeing UFOs. Here's how that conversation went:

> Lovell: "BOGEY AT 10 O'CLOCK HIGH."
>
> Capcom: "This is Houston. Say again 7."
>
> Lovell: "SAID WE HAVE A BOGEY AT 10 O'CLOCK HIGH."
>
> Capcom: "*Gemini 7*, is that the booster or is that an actual sighting?"
>
> Lovell: "WE HAVE SEVERAL. ACTUAL SIGHTINGS."
>
> Capcom: "Estimated distance or size?"
>
> Lovell: "WE ALSO HAVE THE BOOSTER IN SIGHT."

What were these men seeing? Well, if we knew, they wouldn't be UFOs, would they?

My perspective is that there are many geophysical phenomena seen by humans that can't yet be explained. We can't identify their origin. I haven't seen any UFOs during my 53 days in space, but there's no denying that expert observers have seen unidentified objects.

The Least You Need to Know

- Wally Schirra got Project Mercury back on track with a very precise mission: Everything went perfectly.

- Schirra proved that a spacecraft's limited fuel and electricity supplies could be successfully managed to accomplish the mission.

- Gordon Cooper was the first American to spend the night in space (though the sun came up every 90 minutes while he was in orbit!).

- Sleeping in space can present difficulties to astronauts even today; Cooper was a master at it.

From Blecch to Yum: History of Space Food

In This Chapter

- ◆ It's better than it used to be
- ◆ Now available in a store near you
- ◆ Counting calories
- ◆ The glop factor: cohesiveness counts

Let's step away from our history of NASA for a moment to give you a taste of a subject that is still a unique part of the spaceflight experience: food.

The thought of space food used to bring an automatic wince to the face of any astronaut—or cosmonaut, for that matter. There is one thing that anyone who has ever served in the military, eaten in a college cafeteria, or stayed in a hospital can agree on: The food stinks. That used to be the case in space, too. One consolation that we space shuttle astronauts have is that, according to all accounts, space food is a heck of a lot better than it used to be.

Our food aboard the shuttle and ISS in 2002 is certainly better-tasting than it was when I first flew in 1994. When we look at the progress in space food over the 40 years since John Glenn's first American orbits, there's no contest—today's space food is light years ahead.

Of Course, John Glenn Was First

John Glenn was the first American to experiment with eating in space. Scientists weren't sure how the eating and digestion process would be affected by weightless conditions. Would the astronaut be able to swallow in an environment with no gravity?

Glenn tried some applesauce. What he discovered was that eating in space was easy. Swallowing and eating felt normal during weightlessness, and food did not collect in the throat as had been feared.

Glenn found that the biggest problem he had was getting the food *to* his mouth. In response to Glenn's experiments NASA came up with a system to feed the astronauts in space.

Food was packed into aluminum toothpaste-type tubes, and astronauts had to squeeze the food, which was in a semi-liquid form, out of the tube into their mouths.

Other foods were prepared in bite-sized cubes or in a freeze-dried form. This vacuum drying process reduced volume and kept foods from spoiling during storage. Freeze-dried foods are widely available in stores now. So there's another thing we can thank the space program for—or not, if your taste doesn't run to food that's been on the shelf for months.

New and Improved Systems

By the time of the Gemini missions the tubes were phased out. They were scrapped not because they made the astronauts feel as if they were eating Crest, but because they were heavy, the containers weighing more than the food inside.

Space Talk

Here's another non-space vocabulary helper: Food that has been dehydrated by freeze-drying has had all the water taken out of it, so that it will not spoil. To **rehydrate** it means to put the water back into it, so that it is once again edible.

The cubes had their own problems, too. If an astronaut was less than neat and allowed a crumb to escape into the cabin, floating around free, it would have to be chased down immediately, lest it get into the equipment and cause a malfunction. NASA's food scientists solved that problem by coating the cubes in gelatin, making them more cohesive and less likely to release a troublesome crumb.

The freeze-dried foods became easier to eat as well, as new easier-to-use plastic bags were introduced. A squirt gun was used to inject water into the bag to *rehydrate* the food. The astronaut kneaded the outside of the bag with his hands to mix the food up, and then squeezed it into his mouth through a tube.

Dr. Jones's Corner
The Mercury astronauts found the mush in the tube very unpleasant, and also had difficulty squirting the water into the bags to rehydrate the freeze-dried food. Anytime you can't get liquid to go where you want it in free-fall, you've got problems. Ask me about my battle with a grape-drink rain shower aboard shuttle *Endeavour!*

Gourmet Menus

Mercury astronauts were envious when they learned of the expanded menu available to the Gemini crews. While the Mercury astronauts ate what amounted to baby food, the Gemini astronauts could enjoy a meal that included a shrimp cocktail and butterscotch pudding.

The Gemini astronauts had another gourmet advantage over their Mercury predecessors—they had hot and cold water, so that their Tang (which in those days came in one flavor: orange) was no longer the same temperature as their soup.

The Spoon Bowl

By the time the Apollo astronauts flew, astronauts were eating with a spoon for the first time, thanks to an invention called the Spoon Bowl.

The bowl was covered on top, and the top had a zipper in it. The astronaut rehydrated the food inside the bowl. Following rehydration the zipper was opened and the contents could be eaten with a spoon. The food did not fly into the cabin because it was designed to remain cohesive and adhesive. That is, it stuck to itself and to the bowl and spoon.

Cosmic Facts

Gemini astronauts could go four days without having to repeat a meal. In fact, the astronauts were allowed to pick their own meal combinations. Their self-designed diet had to stay within certain nutritional restrictions, though, like providing at least 2,800 calories per day. At least half those calories had to come from carbohydrates, a third from fat, and more than 15 percent from protein.

Skylab: Room for a Dining-Room Table

During 1973–74, the Skylab space station—huge in comparison to the Mercury, Gemini, and Apollo spacecraft—was occupied by three teams of astronauts. The crew would gather around a dining-room table, where their menu offered 72 different food items. And they had a freezer, enabling the astronauts to eat filet mignon and ice cream.

Cosmic Facts

If the food was too dry and started to make a dash for freedom, the astronaut could close the zipper and prevent it from escaping.

Cosmic Facts

Each Skylab astronaut received 4.2 pounds of food per day, including packaging.

The Skylab dinner table consisted of pedestals upon which food trays were mounted. The trays were internally heated, so they served not just to hold the food but to keep it warm. The astronauts did not sit in chairs, but rather used thigh and foot restraints to hold them in a sitting position. When dining, they looked as if they were using invisible chairs.

Pop Tops and Magnets

Skylab's food was contained in aluminum cans with pull-off lids. Astronauts no longer had to drink out of plastic bags, but rather could use plastic bottles that collapsed like an accordian as they emptied.

The floating crumb problem had never been completely solved, and astronauts still had to be careful to move deliberately when handling food to keep it in its container or attached to the spoon.

Bread and Cheese

By the time of the Apollo-Soyuz U.S.-Soviet mission in 1975, astronauts had moved up to fresh bread and cheese. Those astronauts had 80 different types of food to choose from, but in the cramped Apollo cabin, they still lacked the freezer and the self-heating trays that the Skylab astronauts had enjoyed. Instead, the Apollo-Soyuz trio used nonheated trays that attached to their thigh during mealtime using spring and Velcro fasteners. Similar trays are still in use on the space shuttle today.

Today's Space Diner

Space food has really come into its own on the space shuttle and International Space Station. Knives, forks, and spoons aren't magnetized anymore; instead there's a magnetic strip on each meal tray.

Dr. Jones's Corner

To keep silverware from floating around the cabin, all knives, forks, and spoons were magnetized, so they stuck to whatever metal surface one placed them on. The canned and frozen food packages would have created a sizeable pile of trash over the months that astronauts lived aboard Skylab. To manage the waste, the astronauts bagged each meal's trash and popped it through an airlock into an empty fuel tank below their living spaces.

Just about any food is available that can be safely preserved, and packaged inside a container to keep it from floating around the cabin. We use freeze-dried foods, small cans of fruit and pudding, and ready-to-eat, "thermo-stabilized" meals that have been borrowed from the military's array of field rations.

Pasta, stews, shrimp cocktail, hot soup, and even ice cream are on the menu. Packaging is half the battle: Peas might be very annoying floating everywhere, but put them in a sauce and they will cling together and adhere nicely to a spoon.

Planning the Menu

About six months before launch, each shuttle crew gets together with the food specialists at Johnson Space Center for a lunch made exclusively of space food. We sample most of the entrees and side dishes, gulp bite-size snacks, and tuck into the latest recipes rolled out of the food lab. With each food sample, we mark up a grading sheet that assigns each food a taste value between 1 and 10. At the end of the meal, we're in serious need of some zero-g—we leave a couple of pounds heavier!

A few weeks later, armed with our own ratings of the food from the taste test, each astronaut puts together his or her own menu, filling in a meal plan for each day of the mission. We have almost complete freedom of choice, but just to make sure we're getting the nutrition we need, the food specialists recommend supplemental foods that bring the menu into balance. The space walkers, for example, should be eating extra potassium to ensure good heart function during the heavy exertions of an EVA. On a shuttle mission, the meals start to repeat about every four days, but on the International Space Station (ISS), the food planners strive to extend the menu to repeat only once every week or eight days.

Dr. Jones's Corner

Many of us have seen the freeze-dried "Astronaut Ice Cream" for sale at museum shops and NASA visitor centers. While it's great stuff, it's not what we eat in orbit—it's too brittle and crumbly for use in free-fall. Our only space ice cream is just Earth ice cream launched into space in a medical freezer—not too common a treat. But we do get to bring special foods with us. Each astronaut on the shuttle or station can choose one or two taste treats to bring along from Earth. I go in for junk food or treats from my boyhood in Essex, Maryland. On my missions, I've managed to bring along TastyKake chocolate cupcakes and butterscotch krimpets, Hostess Snoballs (bright orange ones for a flight near Halloween), and even some freeze-dried Maryland crab soup. Other crewmembers bring along soft cookies (one astronaut's crumbly toll house cookies were a housekeeping disaster!), fresh bread, or bite-size Swiss chocolate bars—my personal favorite (thanks, Marsha!).

Preparing the Meal

How do we prepare all this good stuff aboard the shuttle and ISS? The first thing you need to know is that space and power limitations keep us from having a refrigerator or freezer aboard. So all food has to be stored, sometimes for months, at room temperature. Before launch, our nutritionists and food technicians pack the array of food and drink packages tightly into the shuttle's mid-deck lockers or into ISS food boxes. On shuttle missions we usually organize the food by meal: Meal B (lunch) of flight day 3, for example.

At mealtime we break out the food packages and "velcro" them to the wall or lockers near the galley. The shuttle galley has a convection oven and hot and cold water available, so we first pop the thermo-stabilized pouches into the oven for heating. Next come the de-hydrated pouches. We slide each one onto a blunt needle in the galley and clamp it in place. Using the galley's control panel, we dial up the amount of water, hot or cold, needed for preparation, and hit the fill button. The galley needle then injects the water through a plastic membrane and tube at the top of the package. If the hot water doesn't heat the food thoroughly, we can pop the rehydrated pouch into the oven with the other packages. After waiting a few minutes to let the water thoroughly moisten the food, we clip the pouches open with scissors. We never even break out the knives and forks—a spoon works on everything you can eat in space.

Drinks are consumed, cold or hot, from squeeze pouches made of aluminum foil. An astronaut chooses a drink pouch from the pantry (the drink powder is already inside) and slides it onto the galley needle. Just dial up the desired water amount and temperature, and hit the fill button. The last step is to insert a straw, complete with a small shutoff valve, through the same tube and membrane used to fill the pouch at the galley. My instant coffee pouches are usually so hot after filling that I can't handle them, so I usually "velcro" them to a locker for a few minutes to cool. On some science missions, the shuttle carries a research freezer intended for returning biomedical samples. On the way to orbit, it's empty, so our food team fills it with ice cream bars. Nothing like a rocket-shaped pop-sicle while racing around the Earth at Mach 25!

Carbonated drinks and freshly perked coffee are challenges that haven't been met yet, but NASA food experts are working on them. NASA once flew competing dispensers of Coke and Pepsi as an orbital test of carbonation, but results were mixed. The drinks weren't really cold, and carbonation in zero-g resulted in a drink pouch full of foam and bubbles— it was hard to get a good gulp of the soft drink itself. As for coffee, any system that could potentially leak boiling liquid to float around the cabin is a safety risk we don't need to take.

Food Stories to Chew On

Every astronaut has some good food stories to share. Following are a few of mine to chew on.

Instead of bread, which generates crumbs and gets stale and moldy quickly on the shuttle, we rely on tortillas for sandwiches in orbit. One of my favorite breakfast menus includes a breakfast burrito. I first squeeze a packet of picante sauce onto a tortilla clipped to my food tray—the sauce serves as the glue for the rest of the ingredients. Then I apply a rehydrated sausage patty, hot from the oven, to the tortilla and sauce. I save the hardest part for last—applying my rehydrated Mexican scrambled eggs, spoonful by spoonful, to the hovering tortilla. Once I was just about to spoon the last of the eggs onto the picante sauce when I carelessly bumped the tortilla from behind—disaster! Eggs and sausage took off across the flight deck in all directions. Grabbing left and right, I popped as many floating egg chunks into my mouth as possible, and recruited my friends to chase down the sausage on the ricochet. Never underestimate the clumsiness of an astronaut still adapting to free-fall. Eating has to be done slowly and carefully.

The foods I have a taste for in orbit seem to change noticeably from my preferences on the ground. After four spaceflights, I've learned to include more spicy foods like shrimp cocktail (helps with stuffy sinuses); in general, foods with more zip help overcome the dulling of the taste buds that I seem to experience in orbit. I add picante or taco sauce to tortilla sandwiches and scrambled eggs. On STS-80, our crew ran out of A-1 steak sauce after we developed the habit of adding it to almost everything we consumed. My last crew left a good sampling of our pantry for the Expedition One crew aboard the ISS, but we were careful to retain a few dozen packets of A-1 to sustain us until re-entry!

Here I am during STS-80 (November-December 1996) carefully spooning Mexican scrambled eggs into a tortilla, using picante sauce to glue together a breakfast burrito. It's easiest to work with these clipped to the blue meal tray; a silver drink pouch is "velcroed" nearby. One false move and I'm done for!

(NASA)

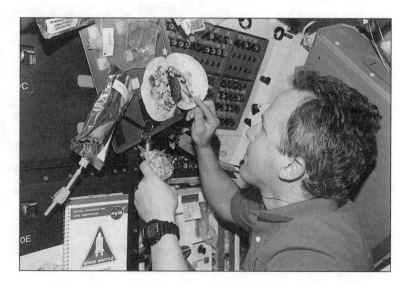

On my last two flights we've managed to launch with some fresh rolls stashed in our fresh food locker. Once everyone's worked up a healthy appetite, we combine the rolls with our Army-derived beef frankfurters, heated up in the galley oven. Add a little mustard, and voilà—hot dogs on fresh rolls! Nothing could be better for a quick lunch in space.

Most crews find that at the end of the day, there's plenty of food left over. Appetites seem to diminish a bit in orbit, perhaps because of the demanding work schedule. My STS-98 crew agreed beforehand that any food left over from the day before was fair game for anyone; we could ignore the little colored dots that identified each package by crewmember and graze freely to supplement our own menus. There's always a lot of horse-trading going on in the shuttle's middeck, as astronauts trade kona coffee for fiesta chicken, or ravioli for another irradiated beefsteak. On my STS-80 flight, the longest shuttle mission ever, we began to exhaust our food supplies as we neared the end of the 18-day flight (extended 2 days due to bad weather at the Cape). We consumed all the spare entrees from the pantry, ate up all the leftovers, and were down to the last few forlorn packages of rice pilaf, bran chex, and Italian vegetables when we headed back to Earth.

On the International Space Station, there's no danger of running out of food, since cargo shipments always maintain a 45-day reserve. The ISS menu borrows heavily from the space shuttle, but it now combines the very best foods from the U.S. and Russian space programs. Russian meals rely more heavily on canned goods than the U.S. menu, but food preparation is similar. The ISS Habitation Module will include a microwave oven, and at least two refrigerator-freezers near the galley, so eventually crews will have food aboard that's close to fresh in terms of texture and taste. Still, nothing like a fresh pizza or a juicy cheeseburger can be found in orbit today.

The final moments of a "chicken flying saucer" sandwich I made aboard Endeavour *during STS-59 in April 1994. I made the sandwich from two tortillas, picante sauce, and an irradiated grilled chicken breast. In space, lunch doesn't get much better than this.*

(NASA)

The Astronaut's Nutritional Needs

Today's space food meets the specific nutritional needs of those who are working in space for extended periods of time. For example, studies have found that astronauts eat less in space than they would on the ground. This can lead to vitamin deficiencies on long missions.

For example, astronauts can develop Vitamin D deficiencies in space for the same reason that winter can affect people on Earth: lack of sunlight.

Ultraviolet light, absorbed through the skin, helps the body produce Vitamin D. But in space, window coatings keep the sun's harsh ultraviolet light from getting to the astronaut's skin, and it's rarely possible to step outside for a walk. We can remedy this through diet: Vitamin D-rich dairy foods (yogurt, cheese, pudding) make it onto the menu.

The bottom line is that while astronauts now get to choose their own menus, they are adjusted so that, when they go into space, they have their basic nutritional needs met. We don't want a deficient diet to exacerbate any of the other difficulties posed by a lengthy stay in space. Astronauts returning from the ISS get a thorough debriefing on their perceptions of the food and their appetites, and medical tests tell us what nutritional needs we've overlooked.

"We have developed a program that helps us ensure that crewmembers go to space with an optimal nutritional status, and that we do everything we can to help them remain healthy while they are there," said Dr. Scott M. Smith of the Johnson Space Center.

The Least You Need to Know

- John Glenn was the first American astronaut to eat in space; the food was close to Mom's apple pie: applesauce.
- The Mercury astronauts' food came out of toothpaste-style tubes—and tasted about as good as you might expect.
- Experience, ingenuity, and new technology have greatly improved the space mealtime experience.
- Today's space food is as good as you can get on any long camping trip. The meals taste better, are more nutritious than they used to be, and can keep a station crew going for months on end.

Two Heads Are Better Than One: Project Gemini

In This Chapter

- ◆ First U.S. two-person crews in space
- ◆ Learning Moon-travel skills
- ◆ Ed White's space walk
- ◆ Cooper's second endurance test

Following Project Mercury, NASA's next manned program was called Project Gemini. Project Gemini was the "tween" program—between Projects Mercury and Apollo.

President Kennedy's lunar goal got NASA working right away on Project Apollo, but the agency soon realized that Mercury would not provide all the know-how to jump off for the Moon anytime soon. The U.S. would need another piloted spacecraft to bridge the gap between the first steps of Mercury and the lunar leap of Apollo. Gemini would be a series of Earth-orbital missions with two-person crews. During these missions, the astronauts practiced the maneuvers (rendezvous and docking) that would later be necessary for lunar visits. To rendezvous and dock, two spacecraft approach one another in space and then link together to form a single orbiting craft.

New Features of the Gemini

To send up a two-person spacecraft, NASA needed a rocket more powerful than the Atlas. It was decided to use the Titan II missile, the largest liquid-fueled intercontinental ballistic missile in the Air Force inventory.

Dr. Jones's Corner

The Titan II rocket had plenty of power to lift the Gemini capsule into orbit. Indeed, it could loft a nine-megaton thermonuclear warhead to nearly any target on the globe. The Titan II had just been deployed to underground silos in Kansas, Arizona, and Arkansas, and the Martin Company took on the job of making the Titan safe enough for astronauts to ride. As a boy, I saw the Gemini-Titan boosters being built at the Martin Company plant in my hometown, just outside Baltimore, Maryland.

The *Gemini* capsule had the same basic shape as the Mercury capsule, a truncated cone with a curved heat shield at the base—but it was much larger. It weighed four tons, as compared to one and a half tons for a Mercury capsule. The two astronauts inside were positioned side by side. Each had his own window.

Space Talk

Project **Gemini** was named after the constellation and sign in the astronomical Zodiac; in mythology, the constellation represents the twins Castor and Pollux.

With input from the Mercury and Gemini astronauts, McDonnell Douglas designed the cockpit to resemble the cockpit of a jet. Each astronaut even had an ejection seat, in case a booster problem called for an emergency escape. (This made the escape tower above the capsule unnecessary.) The other new feature of the Gemini spacecraft was a set of rocket engines at the base, which could be used to change the capsule's orbit in space.

Gemini 2: Last Unmanned Test

There were two unmanned tests, *Gemini 1* and *Gemini 2*. Their objective was to make sure that that the capsule's heat shield would be adequate to protect the spacecraft upon re-entry. The missions were also used to train flight controllers and test the ground communications tracking system.

Twice, in August and September of 1964, the Titan II/Gemini launch vehicle had to be dismantled because a pair of hurricanes (Cleo and Dora) blasted their way across Cape Canaveral. The lift-off of *Gemini 2* was originally scheduled for December 9, 1964, but the Titan II went nowhere—after stage-one ignition, an automatic system detected a loss of hydraulic pressure, causing the engines to shut down approximately one second later. The safety systems, at least, were working.

The *Gemini 2* mission finally got underway on January 19, 1965, at 9:04 A.M. The suborbital mission—similar to the Mercury lobs flown by Alan Shepard and Gus Grissom—ended 18 minutes and 16 seconds later. The spacecraft reached a top altitude of 106 miles and was recovered 2,122 miles downrange. All systems were go so plans went ahead to send two men at once into space aboard *Gemini 3*.

Dr. Jones's Corner

One of the most popular questions I get from people is: "How do you know which way is up or down in space?" Truth is, up and down are wherever you want them to be in space. There's really no way to tell unless you look out the window, but the human mind is a wonderful thing. All you have to do is tell your brain that one part of the spacecraft is the ceiling, and the opposite side is the deck, and almost magically, that's the way your brain will start to organize the information coming in from your eyes.

Gemini 3: The Molly Brown

Alan Shepard was supposed to command the first Gemini manned mission, but before the program got underway, Shepard developed an inner-ear ailment that forced him to be grounded. He was replaced by Gus Grissom, the second American in space.

Accompanying Grissom would be Lieutenant John W. Young, a Navy test pilot with a B.S. in Aeronautical Engineering who had been part of the second group of astronauts selected in September 1962. Poking fun at his first spacecraft, which had sunk, Grissom named the *Gemini 3* spacecraft "Molly Brown," after the popular Broadway musical *The Unsinkable Molly Brown*.

Testing Maneuverability

Gemini-Titan 3 lifted off from pad 19 at Cape Canaveral on March 23, 1965. The mission was to be only five hours long. The objective was to test the spacecraft's operating systems and to determine if the spacecraft's maneuverability was adequate for future rendezvous and docking missions. Molly Brown performed all tasks without a hitch.

The launch of Gemini-Titan 3, with Gus Grissom and John W. Young aboard, on March 23, 1965.

(NASA)

Smuggled Corned Beef

Unhappy with the prospect of space food, Young enlisted the help of a local deli popular with the astronauts and smuggled his provisions aboard the spacecraft.

Astro Bio

John W. Young   John W. Young was born September 24, 1930, in San Francisco, California. Upon graduation from Georgia Tech, Young entered the United States Navy. After serving on the destroyer USS *Laws* for one year, he was sent to flight training and wound up flying Cougars and Crusaders with Fighter Squadron 103 for four years. After completing the course at the U.S. Navy Test Pilot School in 1959, he worked for three years at the Naval Air Test Center at Patuxent Naval Air Station, Maryland. His test projects included evaluations of the Crusader and Phantom weapons systems. In 1962, he set world time-to-climb records to 3,000-meter and 25,000-meter altitudes in the F-4 Phantom II. In September 1962, Young was selected as an astronaut. He is the first person to fly in space six times (seven times if you count his lunar lift-off). His six space missions were *Gemini 3*; *Gemini X*; *Apollo 10*; *Apollo 16*; *STS-1* (first Space Shuttle mission); and *STS-9*.

Today, Young remains an active astronaut at Johnson Space Center. He trains regularly in the shuttle simulator and T-38 jet trainers, and I've had the pleasure of flying with him on many trips to Cape Canaveral. It's one of those can't-quite-believe-it's-true experiences—flying and small talk about space travel with a fellow who's walked on the Moon.

"I was concentrating on our spacecraft's performance," Grissom later said, "when suddenly John asked me, 'You care for a corned beef sandwich, skipper?' If I could have fallen out of my couch, I would have. Sure enough, he was holding an honest-to-john corned beef sandwich."

The astronauts were caught when the spacecraft reeked of corned beef upon recovery and the press got hold of the story. NASA made it clear that there would be no more "carry-on snacks" on future missions.

Dr. Jones's Corner

The Soviets, mindful of their lead in the space race, beat the Americans to the punch when it came to spacewalks, too. Cosmonaut Alexei Leonov aboard *Voskhod 2* left his spaceship for a successful extravehicular activity, or *EVA*. Then, to his horror, he couldn't maneuver back inside.

The pressure in Leonov's suit made it so stiff that he was unable to squeeze back into the airlock. After an anxious few moments, Leonov turned a valve and lowered the pressure in his suit enough so that he could bend. It was a risky gambit, but it worked and he made it back inside. Leonov and his crewmate Pavel Belyayev had a second near-death experience when they overshot their landing zone by 1,000 miles, landed in the mountains, and were almost attacked by wolves in the snowy night.

Gemini IV: These Boots Are Made for Walkin'

Gemini IV (the first Gemini mission to be designated with a Roman numeral), crewed by astronauts James A. McDivitt, commander, and Edward H. White II, pilot, blasted off at a fraction of a second before 10:16 A.M. on June 3, 1965.

Astro Bio

Jim McDivitt   Brigadier General James A. McDivitt, U.S. Air Force, Retired, was born June 10, 1929, in Chicago, Illinois. He earned a B.S. in Aeronautical Engineering in 1949 from the University of Michigan, where he graduated first in his class. McDivitt joined the Air Force in 1951 and flew 145 combat missions during the Korean War in F-80s and F-86s. He is a graduate of the USAF Experimental Test Pilot School and the USAF Aerospace Research Pilot course and served as an experimental test pilot at Edwards Air Force Base, California. He has logged over 5,000 flying hours and was selected as an astronaut by NASA in September 1962. In addition to being the command pilot for *Gemini IV*, he was commander of *Apollo 9*, a 10-day Earth orbital flight launched on March 3, 1969. He retired from the Air Force and left NASA in June 1972, going into private business.

The lift-off followed a flawless countdown. The spacecraft entered Earth orbit at an "insertion speed" of 25,745 feet per second (about 17,500 mph). The apogee of the first orbit was 177.6 statute miles. The perigee was 100.8 miles.

The *Gemini IV* mission, scheduled for four days in length, was to evaluate the effects on man of prolonged spaceflight. Both men—in training for a year for this task—had their abilities to eat, sleep, and maintain a schedule closely monitored while in space.

Space Talk

EVA stands for Extra-vehicular Activity, and it means spacewalking. Any time an astronaut works in space outside his or her spacecraft, it is called an EVA.

Steppin' Out

In addition, White became the first American to step outside of his spacecraft while in orbit, the first American to execute an *EVA*.

The EVA was scheduled to take place on the second orbit, with Astronaut White opening the spacecraft's hatch while over Hawaii, stepping outside the capsule when he was over California, and doing his EVA tasks while he passed over the mainland United States.

On June 3, 1965, Edward H. White II became the first American to walk in space.

(NASA)

Astro Bio

Ed White   Lieutenant Colonel Edward H. White II was born in San Antonio, Texas, on November 14, 1930. He earned a B.S. at the U.S. Military Academy and an M.S. in Aeronautical Engineering from the University of Michigan. He received flight training in Florida and Texas following his graduation from West Point and then spent three-and-a-half years in Germany, flying F-86 and F-100 fighters. In 1959, he attended the Air Force Test Pilot School at Edwards Air Force Base, California, and was later assigned to Wright-Patterson Air Force Base, Ohio, as a test pilot with the Aeronautical Systems Division. He was named an astronaut in September 1962. White died on January 27, 1967, in the *Apollo I* fire.

He was then scheduled to return to his vehicle and close the hatch as they approached the night side of the Earth over the Atlantic.

Special Suit

A special EVA space suit had been built for White, with a protective outer layer, a gold-plated helmet visor, and an attached "life-support pack." The pack contained an emergency bottle of oxygen. Should the oxygen supply from the astronaut's umbilical hose back to the ship fail, there was enough oxygen in the bottle to keep the astronaut alive for eight minutes, enough time for him to return to the interior of the spacecraft.

The suit was made of layers of aluminized mylar, cotton, and felt. The material was designed to protect the astronaut both against extreme temperatures and space particles.

The astronaut wore special over-gloves to protect his hands from the extreme temperatures he might encounter if he should grasp the outside of the spacecraft during his EVA. (Temperatures outside the spacecraft climb to 250°F in the sun and sink to –150°F in the shade.)

Like a lizard with multiple eyelids, the EVA helmet had three visors, rather than the customary one. The fixed inside visor, the one designed to maintain pressure inside the suit, was accompanied by two movable visors. The outer visor was called the sun visor and was easily recognizable by its gold coating. That visor reflected both visible light and infrared rays. Only 10 percent of the sun's visible light was allowed to pass through. The middle visor was a protective layer of polycarbonate plastic.

White was to be attached to his spacecraft by a 25-foot tether. The tether was one assembly composed of three elements: electric wiring, a nylon line, and an umbilical line.

The electrical wiring was important to assure that the space walker could maintain constant communication with his command pilot. The electrical wiring also provided ground control with continuous biomedical readings from sensors on the astronaut's body.

The umbilical line provided a constant supply of oxygen from the spacecraft into the astronaut's suit and created a comfortable breathing environment.

Rendezvous Practice

A very important aspect of Project Gemini was to learn how to rendezvous with other objects in space. Early efforts, including the one on this mission, did not go well.

Astronaut McDivitt had to go to work immediately following the spacecraft's insertion into orbit. He was scheduled to rendezvous and try to station-keep—that is, fly in formation—with the second stage of the Titan II booster, but this experiment was unsuccessful. Trying to scoot closer to the Titan stage was trickier than expected and led to unexpectedly large fuel consumption. The Titan II second stage, which had been scheduled to play a major role in space history, became no more than a footnote. It continued to orbit the Earth for two days before burning up on re-entry over the Atlantic Ocean.

Cosmic Facts

In space rendezvous, you slow down to speed up and speed up to slow down. I'll try to explain: The first rule one has to know is that an object in a low orbit, say 150 miles above the Earth, travels faster than an object in a higher orbit, say 200 miles up. If a low-orbiting spacecraft fires rockets to push itself faster, it also moves into a higher orbit, and moves more slowly around the planet. If a high-orbiting spacecraft fires retro-rockets to slow down, it moves into a lower orbit—and takes less time to orbit the Earth. Therefore, in space, we speed up to slow down and slow down to speed up!

Walk Slightly Delayed

With the countdown to the second-orbit EVA already underway, McDivitt decided not to rush things and delayed White's space walk until the third revolution.

The delay was exactly one orbit, with White opening the hatch over Hawaii and exiting the vehicle as the spacecraft approached the West Coast of the United States.

"I'm out," was White's first message as he became the first American to have nothing between him and outer space but his clothes.

"This is the greatest experience. Just tremendous," White said, while floating in space, 25 feet from the spacecraft, his tether fully extended.

While outside, White used a small handheld maneuvering unit, called a space gun, to move himself around, and found that he could do this with an unexpected degree of accuracy. The gun ran out of compressed nitrogen quickly however, and he was left to spend most of the space walk drifting at the end of his tether.

The Benefits of Training

All good things must come to an end, and White later called his return to the spacecraft "the saddest moment of my life." White stayed outside the spacecraft for 36 minutes and accomplished almost all of what was expected of him. But when he tried to get back into *Gemini IV*, things got a little tense.

As it happened, only days before launch, Commander McDivitt had spent some time practicing taking apart and reassembling the latch that held the spacecraft's hatch in place. Now, as Ed White returned to the capsule, the hatch would not lock back into place (necessary to both repressurize the cabin, and protect the crew from re-entry heat). This potentially fatal malfunction was fixed when McDivitt took the latch apart, reinstalled it, and then locked the hatch back into position. The crew kept the hatch safely closed until after the astronauts returned to Earth almost four days later.

Four Days in Space

Gemini IV returned to Earth on June 7, 1965, four days and two hours after launch. The astronauts had circled the Earth 62 times, almost 186 miles above the planet, and landed 50.5 miles from their scheduled landing site.

The initial plan called for the first computer-controlled re-entry, but this had to be scrapped because of an inadvertent alteration of the computer's memory.

Astro Bio

Charles "Pete" Conrad Jr.   Conrad was born in Philadelphia on June 2, 1930, and earned a degree in Aeronautical Engineering from Princeton University in New Jersey in 1953. Trained as a naval aviator, he became an astronaut in 1962, serving as pilot of the *Gemini V* mission in 1965, the commander of the *Gemini XI* mission in 1966, and commander of *Apollo 12* three years later. In his final space mission in 1973, Conrad commanded the first crew launched to the Skylab space station, which had sustained damage during its launch. During the 28-day mission, Conrad and his crew repaired the station during three harrowing space walks (see Chapter 23). Conrad retired from the U.S. Navy and NASA in 1974. After leaving the space agency, he devoted his energies to developing reusable spacecraft. Conrad died in a July 1999 motorcycle accident not far from his home in Ojai, California.

Gemini V: Endurance Test

The commander of *Gemini V* was Gordon Cooper, the man who had circled the Earth 22 times as a Mercury astronaut (see Chapter 13, "Near Perfection"). His pilot on the mission was Charles "Pete" Conrad Jr. The primary goal of *Gemini V* was to prove that men could live and work in space long enough (about eight days) to go to the Moon and return.

Cooper was NASA's endurance man. He had aced the Mercury endurance test and now, along with Conrad, he was to take on the same challenge for Project Gemini.

The launch of *Gemini V* was originally scheduled for August 19, 1965, but it had to be postponed because of a problem loading the fuel cell that would furnish electrical power. Launched on August 21, 1965, *Gemini V* stayed in orbit for eight days (to be precise, 7 days, 22 hours, 55 minutes, 14 seconds), proving that man could survive and function in space long enough to carry out a lunar landing and return to Earth.

The astronauts also performed experiments leading up to later rendezvous and docking missions. These were the only experiments on the mission that did not go perfectly, due to restrictions caused by a problem with a fuel-cell heater. Because of the reduced electrical power available, the rendezvous work was not attempted.

On the final day, the *Gemini V* crew spoke with Mercury astronaut Scott Carpenter, who was in Sealab II, an undersea station 205 feet underwater off the coast of California, to demonstrate the capabilities of American communication systems.

Cosmic Facts

If you've always wanted to be taller, then becoming an astronaut is the occupation for you. Did you know that astronauts often gain a full two inches in height while in space? That's because the near-weightless conditions cause fluid to gather in the spinal discs, while the spine itself grows slightly longer because it does not have to bear the weight of the body. I grow about an inch and a half on my shuttle flights.

The spacecraft circled the Earth 120 times before re-entry. Once again, Mission Control tried to bring the spacecraft down under computer control. Once again, the system didn't perform as advertised. This time, though, it appeared to be human error, as incorrect navigation coordinates were transmitted to the spacecraft computer from the ground. This caused the capsule to overshoot its landing target by 92 nautical miles. It was almost an hour and a half after splashdown before the crew of *Gemini V* made it onto the recovery ship, the USS *Lake Champlain*.

Not everything had gone perfectly, but the primary goal of the mission had been accomplished with flying colors. We knew humans could live and work in space for long enough for a moon mission, and we had jumped ahead of the Soviets in space endurance capability.

Gemini VI: The Mission That Almost Wasn't

Gemini VI was originally scheduled for late October 1965, with its primary mission being to rendezvous with an unmanned Agena target vehicle. That target vehicle was launched atop an Atlas on October 25, but the booster never made it to orbit. *Gemini VI* was postponed. Juggling the launch order and innovating so that the whole program would not fall behind schedule, NASA sent *Gemini VII* up as scheduled, on December 4, 1965. *Gemini VI* was renamed *Gemini VIa*, and it was rescheduled for a December 12 launch. Instead of rendezvousing with an unmanned target vehicle, it would rendezvous with *Gemini VII*.

On December 12, however, came one of the scariest moments in NASA history. The countdown to launch *Gemini VIa* reached zero, with astronauts Wally Schirra and Thomas Stafford aboard, and the Titan's engines rumbled into life. But, after two seconds of firing, just before lift-off, the Titan II booster shut down. Many observers flashed back to Vanguard's 1957 collapse, fearing the Titan would topple down off the launch pad.

Astro Bio

Thomas Stafford   Lieutenant General Thomas P. Stafford, U.S. Air Force, Retired, was born September 17, 1930, in Weatherford, Oklahoma. He graduated from the U.S. Naval Academy with a bachelor's degree in 1952. Stafford received his pilot wings at Connally Air Force Base, Waco, Texas, in September 1953. He attended the Air Force Experimental Flight Test Pilot School at Edwards Air Force Base and graduated in April 1959. NASA selected him as an astronaut in September 1962. He went into space four times, on *Gemini VI*, *Gemini IX*, *Apollo 10*, and the *Apollo-Soyuz* mission. He retired from NASA in the late 1970s and went into private business and still serves as a NASA advisor.

Controllers fully expected Schirra to command an emergency ejection from the spacecraft, because he had no way of knowing the booster wouldn't slip off the launch stand and explode. (It was later determined that an electrical plug had fallen out of its socket in the booster rocket. The problem was detected by a "malfunction-detection system" which shut down the ignition.) Ejection would have guaranteed the safety of the astronauts, but would have caused the cancellation of the entire mission, a major setback for the Gemini program. Instead, Schirra coolly waited out the problem until it was apparent the booster had safely shut down. Wally Schirra, consummate test pilot, had only burnished his reputation further.

Schirra and Stafford were very lucky that the engine shutdown occurred before lift-off. A piece of equipment called a "launch restraint" held the rocket to the launch pad during the first few seconds of ignition, before releasing and allowing the rocket to lift-off. The restraint was still in place when engine cutoff took place. If the rocket had lifted off even a few centimeters, it would have toppled back onto the pad and exploded. Technicians

Cosmic Facts

The flight plan called for the two spacecraft to close to within a few feet of each other in orbit. Since the Army-Navy football game occurred at the same time *Gemini VIa* and *VII* were in orbit, Annapolis graduates Schirra and Stafford held up a sign in their spacecraft window that said "BEAT ARMY." Astronaut Borman, who had graduated from West Point, replied with a sign of his own, reading "BEAT NAVY." As it turned out, none of them got their wish as the game ended in a 7–7 tie.

quickly fixed the problem, and *Gemini VIa* finally lifted off on December 15. *Gemini VII* had already been in space for 11 days.

Gemini VII: Fortnight in Space

Gemini VII spent 14 days in space with astronauts Frank Borman and James Lovell aboard. The mission proved once again that humans could exist in space for extended periods of time without any serious mental or physical side-effects. Of course, that didn't mean that it was a pleasurable experience.

Since there was no such thing as a space shower in the cramped Gemini cabin, two men sharing the same space for two weeks was a true test of the "Right Stuff." For once, the fact that weightless conditions dull the sense of smell (as discussed in Chapter 13) was a blessing.

Because of *Gemini VIa*'s launch delay, *VII* actually went into space first. These two overlapping missions meant that, for a time, NASA had two manned spacecraft in Earth orbit at the same time.

In the biggest milestone of the mission, the two craft pulled off the first rendezvous in space—thus accomplishing the portion of *Gemini VIa*'s mission that had been put into jeopardy by the failure of the Agena target vehicle. The two craft at one point came within a foot of one another. The crews took beautiful color photographs of each other, so for the first time we got to see what a spacecraft looked like as it orbited the Earth. The successful rendezvous was a huge leap forward on the trail to the Moon.

Astro Bio

Frank Borman   Frank Borman was born on March 14, 1928, in Gary, Indiana, and raised in Tucson, Arizona. He earned a B.S. from the U.S. Military Academy, West Point, in 1950 and an M.S. in Aeronautical Engineering from the California Institute of Technology in 1957. He completed the Harvard Business School's Advanced Management Program in 1970. A career Air Force officer from 1950 on, his assignments included service as a fighter pilot, an operational pilot and instructor, an experimental test pilot, and an assistant professor of Thermodynamics and Fluid Mechanics at West Point. When selected by NASA, Borman was an instructor at the Aerospace Research Pilot School at Edwards Air Force Base, California. He retired from the Air Force in 1970 and later went on to become the CEO of Eastern Airlines.

James Lovell   James A. Lovell was born March 25, 1928 in Cleveland, Ohio. He graduated from the U.S. Naval Academy in 1952 and from the Navy Test Pilot School at Patuxent River, Maryland, in 1958. He was selected to be an astronaut in September 1962.

In addition to *Gemini VII*, he commanded *Gemini XII*, flew aboard *Apollo 8*, and commanded *Apollo 13*. On *Apollo 8*, the first manned mission to orbit the Moon, Lovell, along with fellow crewmembers Frank Borman and William A. Anders, became the first humans to escape the gravitational influence of the Earth. Lovell retired from the Navy and NASA in 1973 and went into private business.

The Gemini VII *and* Earth, *as seen from* Gemini VIa. *The two spacecraft are approximately 43 feet apart.*

(NASA)

The Least You Need to Know

- ◆ Project Gemini flew ten successful two-person spaceflights.
- ◆ Ed White's spectacular EVA was a highlight of the Gemini program.
- ◆ Project Gemini taught us skills we would later need to go to the Moon, such as rendezvous, docking, and long-duration flight.
- ◆ *Gemini VIa* and *Gemini VII* performed the first space rendezvous and flew within a foot of each other more than 100 miles above the Earth.

16

Ironing Out Gemini's Bugs

In This Chapter

- ◆ Neil Armstrong's grace under pressure
- ◆ Learning to dock
- ◆ Exhaustion in space
- ◆ Buzz discovers the EVA key

Following the dual missions of *Gemini VI* and *VII* (see Chapter 15, "Two Heads Are Better Than One: Project Gemini"), NASA scheduled five more Gemini missions as the space agency pursued the knowledge, experience, and technology necessary to walk on the Moon before 1970.

Gemini VIII: Momentary Success and a Very Close Call

With the man who would one day be the first to step on the Moon at the helm, *Gemini VIII* launched into orbit on March 16, 1966. The launch had slipped a day because of problems with the Titan booster hardware.

The commander, Neil A. Armstrong, was making his first trip into space on this flight, with fellow rookie pilot David R. Scott at his side. The mission goal? Chase down and dock with an Agena vehicle, launched hours earlier atop an Atlas rocket.

Astro Bio

Neil Armstrong   The man who would moonwalk into the pages of history was born August 5, 1930, in Wapakoneta, Ohio. He earned a B.S. in Aeronautical Engineering from Purdue and an M.S. in Aerospace Engineering from the University of Southern California. Armstrong flew 78 combat missions during the Korean War. After the war he joined NACA (National Advisory Committee for Aeronautics), NASA's predecessor, as a research pilot at the Lewis Laboratory in Cleveland. He later transferred to the NACA High Speed Flight Station at Edwards Air Force Base, California. He was a project pilot on many pioneering high-speed aircraft, including the 4,000 mph X-15. He has flown over 200 different models of aircraft, including jets, rockets, helicopters, and gliders. He resigned from NASA in 1971 and became a professor of Aerospace Engineering at the University of Cincinnati. In 1979, he left teaching and went into private business.

First Docking

Gemini VIII accomplished its *docking* on the craft's fourth orbit. Docking was essential to pulling off a moon landing. (When putting men on the Moon, the Apollo spacecraft would go into lunar orbit. From that craft would detach a small landing craft which would set down on the lunar surface. The landing craft would then lift off from the Moon, rendezvous with the orbiting craft, and dock for the return trip [see Chapter 18, "Early Apollo Missions"]). So after successfully docking, Armstrong and Scott were feeling on top of the world. But less than half an hour later, a problem with the spacecraft-control system threatened not only the mission, but the lives of the astronauts as well.

Space Talk

Two spaceships have **docked** when they have linked themselves in space.

With the two vehicles linked nose-to-nose, a thruster rocket on the Gemini stuck in the "on" position, and the Gemini capsule and the Agena target vehicle began to yaw and roll uncontrollably. The problem quickly became severe, the astronauts tumbling uncomfortably through space at a rate of one full revolution per second. Had this spinning continued to accelerate, the two vehicles would have been torn apart, or the astronauts would have lost consciousness inside the tumbling cabin.

Anxious Moments

The crew couldn't easily tell that the problem was a stuck attitude-control thruster. They suspected the Agena was the problem, and manually separated their Gemini from the target's docking collar. Because the problem was with the Gemini, this only made the roll problem accelerate. Armstrong fought for some way to stop the roll. The only control system on the spacecraft still working was the re-entry control system, up on the Gemini's nose section.

Acting on instinct, Armstrong used the re-entry system to stop the somersaulting and stabilize the spacecraft. NASA rules stated that any use of the re-entry control system meant the immediate termination of the mission, because that fuel supply and thruster system was essential for returning to Earth. The astronauts prepared for an immediate de-orbit. Splashdown took place in the Pacific, near Okinawa, Japan, which was the secondary recovery site. The emergency return cancelled Scott's scheduled space walk (boy, I can relate!).

Gemini VIII completed only seven orbits and was in space for 10 hours, 41 minutes, and 26 seconds before splashing down in the Pacific.

> **Dr. Jones's Corner**
>
> Because of the emergency return to Earth and the remoteness of the splashdown site, three hours elapsed after landing before the *Gemini VIII* astronauts were hauled safely aboard the recovery ship, the USS *Guam*.

> **Astro Bio**
>
> **David Scott**   David R. Scott was born June 6, 1932, in San Antonio, Texas. He received a B.S. from the U.S. Military Academy in 1954, standing fifth in a class of 633, and an M.S. in Aeronautics and Astronautics from the Massachusetts Institute of Technology in 1962. As a NASA astronaut, Scott flew on *Gemini VIII* and *Apollo 9* and was commander of *Apollo 15*. He retired from the U.S. Air Force in March 1975 with the rank of Colonel and over 5,600 hours of flying time.

Gemini IX: A Tragic Revamping

Gemini IX was originally scheduled to go up in May 1966, with the flight plan again calling for the spacecraft to rendezvous and dock with an Agena target vehicle. Once again, events did not unfold as planned, and NASA showed its versatility by revamping the mission to fit existing conditions.

Once again, an Atlas-Agena malfunction prevented the target vehicle from making it into orbit, and *Gemini IX* became *Gemini IXa*, with a launch date set in June. On June 1, an "Augmented Target Docking Adapter," an off-the-shelf substitute for the Agena, was put into orbit by an Atlas rocket. The schedule called for *Gemini IXa* to go into orbit that same day, but a computer guidance system problem caused a two-day delay.

The mission, for tragic reasons, also did not fly with its original crew. Those men, Elliott M. See and Charles Bassett, were killed in a NASA T-38 jet trainer crash on February 28, 1966. The backup crew moved up to the prime crew positions. The new crew comprised commander Thomas P. Stafford and pilot Eugene A. Cernan. Seemingly jinxed, *Gemini IXa* finally blasted off into space at 8:39 on the morning of June 3, 1966.

An "Angry Alligator"

The rendezvous and docking, scheduled to take place during the mission's third orbit, was only partially successful. When the crew pulled up alongside the target adapter, a shroud still partially covered the "female" portion of the docking cone. The open-jawed shroud, which had failed to jettison after launch, reminded Stafford of "an angry alligator." Now ad-libbing under the changed circumstances, the crew performed a series of rendezvous maneuvers, including a run-through of thruster firings that would be needed to orbit and land on the Moon. The other feature of the mission was a space walk by Cernan, and it proved to be a near-fatal affair.

Astro Bio

Eugene Cernan   Eugene A. Cernan was born in Chicago, Illinois, on March 14, 1934. He received a B.S. in Electrical Engineering from Purdue, and an M.S. in Aeronautical Engineering from the U.S. Naval Postgraduate School, Monterey, California. Commissioned through the Navy ROTC Program at Purdue, he entered flight training upon graduation and was assigned to Attack Squadrons 26 and 113 at the Miramar, California, Naval Air Station. Cernan was one of 14 astronauts selected by NASA in October 1963. His missions included *Gemini IX*, *Apollo 10*, and *Apollo 17*. Cernan was the second American to walk in space, one of the three men (the other two were Lovell and Young) to have flown to the Moon twice, and was the last man to leave his footprints on the Moon, as commander of *Apollo 17*. On July 1, 1976, Captain Cernan retired after more than 20 years with the U.S. Navy and went on to start his own company.

Spacewalkin' Blues

One painful lesson NASA learned from Cernan's walk was that it took more energy than originally thought to work in a spacesuit. Although his EVA lasted two hours and seven minutes, it was cut short because Cernan was overheating inside his space suit, and his visor was almost completely fogged over.

The potentially fatal problem, it turned out, was caused by the fact that Cernan's every move had to work against the *internal pressurization* of his space suit.

The suit did not want to bend. His body also turned inadvertently away from his work site with even the slightest move—Newton's Third Law once again. His heartbeat was up to 180 beats per minute by the time he was called back into the spacecraft. Exhausted and barely able to see, Cernan struggled painfully to force himself back down into his seat and close the hatch above him. Tom Stafford even ran through in his mind the grim possibility of having to cut the umbilical, which would have been necessary to close the hatch, and leave Cernan adrift in orbit. Cernan's ordeal was a wake-up call for NASA's EVA planning effort.

Gemini IXa was in space for almost three days, orbiting the Earth 45 times. Re-entry and splash-down went perfectly, and the crew missed the target by less than a mile.

> **Space Talk**
>
> The **internal pressurization** is the amount of oxygen that has been pumped into the space suit. Imagine a bicycle tire that has had too much air pumped into it. Because of its excessive internal pressurization, the tire is rock hard and inflexible. Cernan's suit pressure was only about four pounds per square inch (psi), but that was enough to make the suit swell like a balloon.

Dual Rendezvous: *Gemini X*

Gemini X, with John W. Young and Michael Collins aboard, rocketed into orbit on July 18, 1966, at 5:20 P.M. This time, everything came together—the spacecraft successfully rendezvoused and docked with the Agena upper stage. The *Gemini X* crew accomplished the feat on their fourth revolution around the Earth.

Astro Bio
Michael Collins   Brigadier General Michael Collins (U.S. Air Force Reserve, retired) was born on Halloween, 1930, in Rome, Italy. He received his B.S. from the U.S. Military Academy at West Point in 1952. Following graduation, Collins chose an Air Force career and served as a test pilot for the Air Force Flight Test Center, Edwards Air Force Base in California. After logging approximately 5,000 hours of flying time, Collins was chosen among the third group of astronauts selected by NASA in 1963. After serving as a member of the backup crew for *Gemini VII*, Collins was named the pilot for *Gemini X*. Collins's performance on that mission was among the best ever, as he and Young fulfilled nearly every goal in the flight plan, even catching up later with the old *Gemini VIII* Agena. His sterling record got him the slot as command module pilot for *Apollo 11*, the mission that first landed men on the Moon. In total, Collins logged 266 hours in space. One hour and 27 minutes of that was spent on his *Gemini X* space walk. Collins left NASA in January 1970 and later served as Director of the National Air and Space Museum, Smithsonian Institution, in Washington, D.C. In 1974, Collins wrote the book *Carrying the Fire* (Farrar, Straus & Giroux), a critically acclaimed account of his *Apollo 11* experiences.

The mission also set a new human altitude record when the crew used the Agena engine to climb to an orbit 475 miles above the Earth. The record did not last very long, however, for it was broken only eight weeks later by the crew of *Gemini XI*.

The flight included a 1 hour, 29 minute EVA by Collins, which went a long way toward overcoming the problems Cernan had encountered outside *Gemini IX*. The crew was in space for 2 days, 22 hours, 46 minutes, and 39 seconds, completing 43 orbits of the Earth. All the mission's primary objectives were met, although limited fuel reserves led to cancellation of some secondary experiments.

Gemini X splashed down 2.6 miles from the USS *Guadalcanal*, and became the second spacecraft in the Gemini program to land within visual and camera range of the prime recovery ship.

Cosmic Facts

The *Gemini XI* mission had been postponed twice, on September 9 and 10, once because of a fuel leak in the Titan rocket's first stage, and once because of a problem with the auto-pilot.

Cosmic Facts

As the *Gemini XI* spacecraft rose farther and farther from our world, heading toward its altitude record, Pete Conrad could be heard saying, "It's going, it's really going! Whoop-de-do! ... It's fantastic! You wouldn't believe it! I've got India in the left window. Borneo under our nose, and [Australia is] right at the right window. The world is round!"

First Tethered Flight: *Gemini XI*

Closing in on the final demonstrations of the Apollo mission requirements, *Gemini XI*, carrying astronauts Pete Conrad and Richard F. Gordon Jr., raced into orbit on September 12, 1966. They were hot in pursuit of a Gemini-Agena target vehicle launched into orbit earlier the same day.

A New High: 850 Miles

After linking up with the Agena, the crew of *Gemini XI* set a new human altitude mark, blowing away the record set only eight weeks before by *Gemini X*. The crew soared to a dizzying 850 miles.

The Gemini capsule continued to work with the Agena vehicle; during an EVA, astronaut Gordon tied a tether between the two vehicles with a 30-meter (nearly 100 foot) line.

Rocket Rodeo

One of the most memorable moments of the mission came during Gordon's EVA (extravehicular activity, space walk). Before tethering the Agena target vehicle to the Gemini spacecraft, Gordon found himself straddling the cylindrical fuselage of the target rocket. Seeing what his partner was doing, Conrad called out: "Ride 'em, Cowboy!"

The *Gemini XI* spacecraft orbited the Earth 44 times, staying aloft for just 42 minutes shy of three full days. The mission tested a completely automated re-entry system on September 15, which worked perfectly.

The spacecraft missed its landing site by less than three miles. Conrad and Gordon were aboard the recovery ship, the USS *Guam*, only 24 minutes after splashdown.

Grand Finale: *Gemini XII*'s Five-Hour Space Walk

The final Gemini flight, *Gemini XII*, was launched during the mid-afternoon of November 11, 1966, with commander James A. Lovell and pilot Edwin E. "Buzz" Aldrin aboard. On this last mission of the series, the team accomplished all of its primary objectives, and NASA came out ready to start the Apollo program.

The capsule rendezvoused and docked with the Agena target vehicle on the third revolution. Aldrin's EVA was the centerpiece of the mission. Learning from Eugene Cernan's experience, Aldrin conserved his energy and was able to work outside the spacecraft for five and a half hours.

One of the reasons Aldrin was able to work so efficiently outside his spacecraft was that he had trained for his EVA by working underwater in his space suit. This technique, he found, closely simulated the weightless conditions in space. His success made underwater EVA training essential for all future spacewalkers.

The *Gemini XII* capsule orbited the Earth 59 times and splashed down less than three miles from its target. It was a fitting finale to the Gemini program, which had catapulted the U.S. well ahead of the Soviets in the race to the moon.

> ### Dr. Jones's Corner
>
> Lovell and Aldrin had done their jobs spectacularly well, and both were rewarded with Apollo missions. Aldrin, of course, was the second man to walk on the moon, while Lovell flew on the first lunar orbit mission, *Apollo 8*, and commanded the ill-fated flight of *Apollo 13*.

> ### Cosmic Facts
>
> The NASA budget peaked in 1966: $5.9 billion. That was close to one percent of the Gross National Product. During that year there were 36,169 NASA employees. NASA's contractors had a labor force 411,000 strong. It was estimated that more than one million Americans lived in households that received their income from the space program.

Astro Bio

Buzz Aldrin   Retired Air Force Colonel Buzz Aldrin, Ph.D., was born January 20, 1930, in Montclair, New Jersey. He attended West Point, where he graduated third in his class in 1951. As a U.S. Air Force pilot, Aldrin flew 66 combat missions over Korea and is credited with shooting down two MiG-15 aircraft. He subsequently earned his doctorate from MIT. His thesis turned out to be useful in his future career; it was titled "Guidance for Manned Orbital Rendezvous."

In October 1963, Aldrin was named one of the third group of astronauts. His first trip into space was aboard *Gemini XII* in November 1966, the final flight in that program. He is best known for his second flight into space, as the lunar module pilot for *Apollo 11*. On July 20, 1969, Aldrin became the second man to walk on the Moon.

After spending almost 290 hours in space, almost eight hours of that outside his spacecraft, Aldrin resigned from NASA in July 1971. He went on to write two non-fiction books, *Return to Earth*, an autobiography, and *Men from Earth*, about the Apollo program. He lectures around the world, has published a science fiction novel, and is today President of Starcraft Enterprises, Laguna Beach, California. Aldrin is still a leading space advocate. In June 2001, he testified before Congress in favor of encouraging the budding space tourist industry.

Bodily Changes

One of Project Gemini's main objectives was to find out how lengthy stays in space affect an astronaut's health. What NASA learned was that while physical changes do occur during long stretches in space, none of them were serious enough to threaten a Moon mission.

Doctors found that astronauts returning from space had lost a small amount of bone density. The Gemini astronauts also lost some calcium and protein nitrogen, but again only in very small quantities.

Red blood cells decreased by between five and 20 percent, and the heart was slightly weakened by an extended stay in weightless conditions. These conditions all quickly corrected themselves after the astronaut's return to Earth. Later, we found that exercise in space and dietary supplements further minimized these effects.

Some physical functions were entirely unaffected by long stays in space. Blood pressure, coordination, and muscle tone remained unchanged.

And that was it for the Gemini program. The total budget for Project Gemini had been $1.15 billion. While no one assumed it would be easy, NASA was cautiously optimistic about its chances for meeting Kennedy's goal. Next up for America—Apollo … and the Moon!

The Least You Need to Know

- ◆ On Neil Armstrong's first mission, he deftly handled an emergency in space.
- ◆ Eugene Cernan's space walk was cut short due to an exhausting workload that nearly proved fatal.
- ◆ Buzz Aldrin successfully trained for his EVA by simulating the weightless environment underwater.
- ◆ Two-week stays in space—the length of an extended lunar mission—do not affect coordination or muscle tone.

17

Scouting the Solar System

In This Chapter

- ◆ Martian mission
- ◆ Venusian investigation
- ◆ Mercury rising
- ◆ Lunar TV

While the Mercury and Gemini projects were preparing NASA for an eventual manned mission to the Moon, the unmanned portion of our space program was busy as well. Robotic missions in the 1960s not only did their part in gathering data about the Moon for the Apollo engineers, but also, using state-of-the-art photographic equipment, were taking us places no human could go, revealing sights no human could ever imagine.

Mariner

In Chapter 9, "Explorer and Pioneer," I told you about Project Pioneer, which was investigating the Sun and interplanetary space during the 1960s—measuring the solar wind, cosmic rays, and the solar magnetic field. But Pioneer was not the only program reaching outward into the solar system, looking beyond the Moon to the planets and the Sun. The program designed to investigate our nearest planetary neighbors was called Mariner.

The first of the Mariner series of spacecraft to head outward into the solar system was *Mariner 2*, which had Venus, the second planet from the sun, as its target. The 446-pound *Mariner 2* spacecraft flew past Venus on December 14, 1962. The most important discovery of the flight was that the temperature on the surface of the planet was 800 degrees Fahrenheit, which was just about twice as hot as had been predicted. Prior to *Mariner 2*'s *flyby*, science fiction writers had depicted Venus as a hot, rainy, perpetually cloudy planet, but the new results made it clear that the Venusian surface was a blistering furnace.

> **Space Talk**
>
> A **flyby** is when a space probe flies close enough to a planet to photograph it and gather other information, but not so close that it is captured into orbit by the planet's gravitational pull.

> **Space Talk**
>
> **Canali** is Italian for channels, or grooves. It can also translate as canal, a man-made waterway built for transportation purposes. You can't have canals on a world that has never known life—but more than a few scientists and science-fiction writers were captivated by the latter possibility.

Red-Planet Reconnaissance

Even during the early days of astronomy, observers' eyes were often drawn to the red beacon of Mars. Adding intrigue to our fascination, in 1877, Italian astronomer Giovanni Virginio Schiaparelli spotted through his telescope dark lines criss-crossing the Martian surface. He called these lines *canali*.

We have since learned that these were not canals, but merely optical illusions caused by low-resolution telescopes. But the possibility of life on Mars remained a primary question among the many mysteries of space. NASA was as curious about Mars as anyone on Earth, so only weeks after launching *Mariner 2* toward Venus, NASA sent out twin missions, *Mariners 3* and *4*. Their destination: the Red Planet itself.

Mariners 3 and 4

Mariner 3 was launched on November 5, 1964, atop an Atlas Agena D rocket. The mission was a failure, however, when the launch vehicle's nose fairing failed to jettison, trapping the spacecraft inside. (The nose fairing is a smooth, aerodynamic covering for the tip of a rocket.)

Mariner 4, a product of NASA's "always have a backup" philosophy, launched three weeks later, on November 28, 1964. This time the Atlas-Agena functioned smoothly, and eight months later, *Mariner 4* became the first spacecraft to take close-up pictures of Mars.

The satellite carried special solar panels designed to absorb solar energy; the craft was oriented during flight so that the panels faced the Sun at all times.

Pictures of Mars

It took *Mariner 4* 228 days to get to Mars. The satellite passed within 6,200 miles of the Martian surface on July 15, 1964. As it passed, it took 22 TV pictures that documented about one percent of the Martian surface. Instead of a network of canals, *Mariner 4* photographed a pitted desert, a vast wasteland of red sand and craters. Mars appeared to be about as sterile as the Moon's blasted surface. But *Mariner 4*'s mission wasn't over once it passed Mars. The satellite looped around the Sun in a huge solar orbit and passed near the Earth again in 1967. Never ones to waste an opportunity, NASA technicians used the satellite to test technologies that would later be important on future interplanetary missions. *Mariner 4* ceased operations on December 20, 1967.

Dr. Jones's Corner

The Atlas/Agena rocket, which was used to launch the Mariner probes, was a multipurpose two-stage liquid-fuel rocket, used to place unmanned spacecraft in Earth orbit. The Atlas/Agena D stood 100.4 feet high. It developed a total thrust at lift-off of approximately 376,000 pounds. The Agena upper stage was also used as the target docking vehicle for the Gemini rendezvous missions.

Cosmic Facts

Programs that used the Atlas/Agena rocket included:

- Early Mariner probes to Mars and Venus
- Ranger photographic missions to the Moon
- The Orbiting Astronomical Observatory
- Early Applications Technology Satellites
- Project Gemini, as a rendezvous target vehicle
- The Lunar Orbiter, five of which went into orbit around the Moon and mapped most of the lunar surface in preparation for the manned lunar landings

Venusian Voyage: *Mariner 5*

Mariner 5 was launched on June 14, 1967, atop an Atlas/Agena D rocket. The octagon-shaped satellite was propelled into orbit around the Sun. Part of that orbit would involve a close flyby of Venus. The satellite carried several experiments. Venus's atmosphere was bombarded with radio waves and scanned for the brightness of its reflected ultraviolet light, which would reveal cloud formations. Samples of solar particles were taken, and the magnetic field fluctuations above Venus's surface were measured.

Mariner 5 came closest to Venus, less than 2,500 miles away, on October 19, 1967, and was last heard from in November of that year. It confirmed the hellish conditions on the Venusian surface—crushing pressures 90 times that of Earth and temperatures of 900°F, hotter than a self-cleaning oven.

Twin Missions to Mars: *Mariners 6* and *7*

Mariner 6 was launched atop an Atlas/Centaur rocket on February 24, 1969. Its sister mission, *Mariner 7*, was sent on the way to Mars by an identical rocket three days later. The eight-sided, 11-foot satellites traveled in tandem to Mars where they were to photograph the Martian surface. Each had a small rocket engine in case changes in course needed to be made. The probes carried instruments to measure the amount of infrared and ultraviolet light reflected from Mars' surface. They also studied the chemical make-up of the Martian atmosphere. Each had two television cameras.

Mariner 6 flew by the planet on July 31, 1969. *Mariner 7* flew by on August 4. *Mariner 6* ended up returning 75 pictures while *Mariner 7* took 126. Photographs were taken of both of Mars's polar ice caps. Photos were also taken of Phobos, one of Mars's two tiny moons. The additional images seemed to reinforce *Mariner 4*'s conclusion that Mars was a cold, battered, lifeless world. Total cost of the two missions: $148 million.

First Man-Made Object to Orbit Another Planet: *Mariner 9*

On the next U.S. attempt on Mars in May of 1971, a failure with the Centaur upper stage dumped *Mariner 8* into the ocean. (One wag at NASA suggested *Mariner 8* be redesignated an oceanographic research satellite.) Our luck turned when *Mariner 9* was launched successfully on May 30 of that year. This time everything worked properly, and *Mariner 9* became the first man-made object to go into orbit around another planet. On November 13, 1971, *Mariner 9* entered Martian orbit and began to photograph the Martian surface and analyze the Martian atmosphere with its infrared and ultraviolet equipment.

At the time *Mariner* arrived, the Martian weather was atrocious, and much of the surface was obscured by massive dust storms. Two Russian landers went straight into the storms and crashed; it was decided that *Mariner 9* would wait for the weather to clear. *Mariner 9*'s first stunning discovery came during these storms. Although the clouds of dust were many miles high, the spacecraft observed a mountain that was sticking up above the haze. It turned out to be a volcano, since named Olympus Mons (Mount Olympus). It is the largest mountain on Mars and the largest known volcano in the solar system, standing 18 miles high and measuring 373 miles across, the size of the state of Arizona.

It took a couple of months for the storms to diminish; after that the photographs transmitted back to Earth exceeded all expectations. One hundred percent of the Martian surface was mapped by *Mariner 9*, and the craft also took sharp photographs of Mars's two moons, Phobos and Deimos. *Mariner 9* continued mapping for 349 days, and before it was turned off, it had returned 7,329 images to Earth.

> **Cosmic Facts**
>
> *Mariner 9* was shaped like an octagon and was made of fiberglass, aluminum, and magnesium. Solar panels allowed the craft to produce its own power—800 watts while orbiting the Earth, and 500 watts when orbiting Mars.

A List of Firsts: *Mariner 10* Heads Sunward

Mariner 10, launched November 3, 1973 atop an Atlas-Centaur booster, was destined to rack up a list of "firsts." It was the first ...

- ◆ Spacecraft to fly by two different planets.
- ◆ Mission to use the gravitational pull of one planet to reach another.
- ◆ And the only mission to date to take close-up photographs of Mercury.

Mariner 10 flew by Venus on February 5, 1974, and then made three passes by Mercury, on March 29, 1974; September 21, 1974; and March 16, 1975.

Up until *Mariner 10*'s photographic efforts—it took more than 4,000 photos of the planet—scientists were not even certain what the precise shape of Venus was. They discovered that it was nearly round—the images also revealed the global motion of the scalding layers of clouds shrouding Venus.

Mariner 10 came within 469 miles of Mercury and took the only close-up photos of the closest planet to the Sun. The photos showed that Mercury was heavily cratered, like our Moon; *Mariner 10* also detected a large metallic core, and just the faintest trace of atmosphere, composed mostly of helium.

After its close pass by Mercury, *Mariner 10*'s solar orbit carried the spacecraft by Mercury two more

> **Cosmic Facts**
>
> *Mariner 10* was turned off on March 24, 1975, after its fuel supply ran out. It was so tough getting to Mercury that another spacecraft won't head that way until nearly 2010.

times. The second flyby allowed *Mariner 10* to photograph the sunlit side of the planet again as well as the planet's South Pole. On the third flyby the spacecraft measured Mercury's magnetic fields and took data on particles in the environment.

Ranger: Live from the Moon

The Ranger project was the baby of the Jet Propulsion Laboratory. The idea was to launch a probe with a television camera on it toward the Moon. The spacecraft would crash into the Moon, transmitting photos of the lunar surface back to Earth the entire way.

After what seemed like an endless string of failures, Ranger finally got around to doing what it was supposed to do. The first two Ranger missions, launched during 1961, failed to get out of Earth orbit when their Agena second stages didn't ignite.

Ranger 3, which headed moonward in January 1962, did get out of Earth orbit, but missed its target when the craft's main guidance system failed. The satellite was going to miss the Moon by an estimated 20,000 miles when NASA technicians successfully executed a mid-course correction, firing Ranger's onboard engine via remote control. Still not close enough to hit the Moon's surface, the craft's onboard computer then refused to work, and no photographs could be taken or transmitted.

Ranger 4 looked to be a success at first, as its Atlas-Agena rocket put it on a course to hit the Moon, but in the long run it too was a disappointment. The craft's onboard timer failed, so it never deployed its solar panels or radio antenna. Mute and starved of energy, it crashed uselessly into the lunar farside.

Ranger 5 was launched in October 1962, but a series of short circuits onboard sent it spinning past the Moon into solar orbit. Following *Ranger 5* the craft received a major re-design. *Ranger 6*, which launched in January 1964, was a simpler but more reliable space-craft. It launched perfectly, executed its mid-course correction perfectly and crashed into the Moon's Sea of Tranquility just as it was supposed to. There was only one problem. Its TV cameras didn't work, and the spacecraft slammed into the lunar surface without returning a single image.

Then came *Ranger 7*, and finally, success. Launched in July 1964, *Ranger 7* transmitted 4,316 photographs of the Moon successfully back to Earth right up until the moment it crashed into the Mare Nubium sector of the Moon. There were two more Ranger missions, both flown early in 1965, and they, too, worked. The final Ranger transmitted its photos directly to the U.S.'s television networks, which showed them on live TV with the words "LIVE FROM THE MOON" superimposed on the screen. The long series of Ranger failures had provided valuable experience to spacecraft engineers, and ultimately gave us our first glimpses of the details of the Moon's surface.

Lunar Orbiters and Surveyors

During 1966–67, five spacecraft known as Lunar Orbiters were sent to the Moon as photographic scouts. The primary goal of the project was to choose safe and interesting landing sites for the upcoming Apollo missions. All five worked as planned, and 99 percent of the Moon's surface was photographed.

Cosmic Facts

Here are the pertinent facts for the five Lunar Orbiter missions:

- *Lunar Orbiter 1*: Launched August 10, 1966; photographed Moon August 18–29, 1966.
- *Lunar Orbiter 2*: Launched November 6, 1966; photographed Moon November 18–25, 1966.
- *Lunar Orbiter 3*: Launched February 5, 1967; photographed Moon February 15–23, 1967.
- *Lunar Orbiter 4*: Launched May 4, 1967; photographed Moon May 11–26, 1967.
- *Lunar Orbiter 5*: Launched August 1, 1967; photographed Moon August 6–18, 1967.

The Lunar Orbiters had an imaging system that consisted of a dual-lens camera, a film-processing unit, a readout scanner, and a film-handling apparatus. Like the Corona Earth-orbiting reconnaissance satellites, the Lunar Orbiters captured their images on film and processed it onboard. Then they scanned the film and "faxed" the images back to Earth.

The Surveyor missions, launched between 1966–68, were the first U.S. spacecraft to land on the Moon. The main objectives of the Surveyors were to make soft-landings on the Moon, obtain close-up images of the surface, and determine if the terrain was safe for manned landings. Each Surveyor was equipped with a television camera. In addition, *Surveyors 3* and 7 each carried a small scoop that dug trenches and tested the lunar soil. Several also had the capability to determine the chemical elements present in the lunar soil.

There were seven Surveyors, five of which worked beautifully. *Surveyor 2* crashed on the Moon, and *Surveyor 4* lost radio contact two and a half minutes before it was to land on the Moon. All the others touched down gently and took stock of their stark lunar surroundings. The thousands of images and scientific data radioed home led NASA to conclude that an Apollo lander could safely touch down and take off again from the Moon.

Viking: Exploring the Surface of Mars

Just a few years after the Apollo landings succeeded, the U.S. set its sights on putting a robot probe down on the surface of Mars. The successor to the Mariner Mars probes was called Viking. At a cost of slightly less than $1 billion, NASA landed two robotic space-craft on Mars and put another two into orbit around the Red Planet. The twin missions mapped the surface in startling clarity and began the search for life on another world.

The spacecraft being sent to Mars weighed just shy of four tons. *Viking 1* blasted off from the Kennedy Space Center on August 20, 1975. *Viking 2* took off three weeks later on September 9. After scouting from Mars orbit for a safe landing site, *Viking 1* landed in an area called the *Chryse Planitia*, on July 20, 1975. Two weeks later, *Viking 2* set down at *Utopia Planitia*, which was on the other side of the planet. Though the photographs sent back of Mars's barren red landscape were stunning, the primary goal of the lander mission was to hunt for evidence of life on Mars. Despite some tantalizing readings from the instruments aboard, the biology experiment turned up no proof that life existed on Mars.

Cosmic Facts

The Jet Propulsion Laboratory built the Viking orbiters, while NASA's Langley Research Center tackled the landers and supervised the entire project.

The rocky Martian land-scape, as seen from Viking 2.

(NASA)

Dr. Jones's Corner

Dr. Joseph Miller, an associate professor in the University of California's department of cell and neurobiology, says he has interpreted information gathered by one of the Viking landers in 1976 to conclude that life might have existed, at least long ago, on Mars. Most scientists did not agree with Dr. Miller's findings, however, concluding that chemical, rather than biological, processes produced the data sent back by the lander. But the discussion on the biology results is still active, and the Viking orbiter images are still used to choose landing sites for our latest generation of Mars probes. NASA never throws away good data, and the fact that information gathered 25 years ago is still the subject of scientific debate strikes me as a good thing.

Voyager

The twin spacecraft *Voyager 1* and *Voyager 2* were launched by NASA in separate months in the summer of 1977 from Cape Canaveral, Florida. As originally designed, the Voyagers were to conduct close-up studies of Jupiter and Saturn, Saturn's rings, and the larger moons of the two planets.

To accomplish their two-planet mission, the spacecraft were built to last five years. But as the mission went on, and with the successful achievement of all its objectives, the additional flybys of the two outermost giant planets, Uranus and Neptune, proved possible—and irresistible to mission scientists and engineers at the Voyagers' home at the Jet Propulsion Laboratory in Pasadena, California. As the spacecraft flew across the solar system, remote-control reprogramming was used to endow the Voyagers with greater capabilities than they possessed when they left the Earth. Their two-planet mission became four. Their five-year lifetimes stretched to 12 and more.

The Voyager mission was designed to take advantage of a rare geometric arrangement of the outer planets in the late 1970s and the 1980s that allowed for a four-planet tour with a minimum of propellant and trip time. This layout of Jupiter, Saturn, Uranus, and Neptune, which occurs about every 175 years, allows a spacecraft on a particular flight path to swing from one planet to the next without the need for large onboard propulsion systems. The flyby of each planet bends the spacecraft's flight path and increases its velocity enough to deliver it to the next destination. Using this "gravity assist" technique, first demonstrated with NASA's Mariner 10 Venus/Mercury mission in 1973–74, the flight time to Neptune was reduced from 30 years to 12.

Eventually, between them, *Voyager 1* and *2* would explore all the giant outer planets of our solar system, 48 of their moons, and the unique systems of rings and magnetic fields those planets possess. Had the Voyager mission ended after the Jupiter and Saturn flybys alone, it still would have provided the material to rewrite astronomy textbooks. But having doubled their already ambitious itineraries, the Voyagers returned to Earth information over the years that has revolutionized the science of planetary astronomy, helping to resolve key questions while raising intriguing new ones about the origin and evolution of the planets in our solar system.

The Least You Need to Know

- *Mariner 4* took the first close-up photographs of Mars.
- *Mariner* missions did the early scouting of both Mars and Venus.
- *Mariner 10* is the only spacecraft ever to visit, and photograph, the planet Mercury.
- *Ranger 7* transmitted the first live close-up television pictures before crashing into the Moon.
- The *Viking* landers touched down on Mars and sent back thousands of photos and valuable weather data, but found no traces of life.

Part 4

Moonwalkin'

Now we turn to the story of Apollo, the series of missions that sent 12 Americans to walk on the Moon, and 12 others to circle our natural satellite. Hundreds of years from now, this achievement will, I believe, be the event for which the twentieth century is best remembered. Nevertheless, the Apollo program could not have gotten off to a worse start. We'll begin with the tragic *Apollo 1* fire that killed three astronauts and show you the resolve that allowed the program to recover from the tragedy and beat the deadline set by President Kennedy eight years before.

After a quick examination of the hardware that NASA developed to fly humans to the Moon, we'll look at each Apollo mission and follow Neil Armstrong, Buzz Aldrin, and Michael Collins as they take *Apollo 11*—and humanity—to the surface of the Moon for the first time.

Chapter 18

Early Apollo Missions

In This Chapter

- *Apollo 1* tragedy
- Safety changes made
- Man orbits the Moon
- Christmas Eve prayers

The robot Rangers, Surveyors, and Lunar Orbiters had scouted the Moon for safe landing sites. The Gemini crews had demonstrated the skills that would be needed to get to and from the lunar surface. The stage was set for Apollo—the moon landing program set in motion by President Kennedy's challenge to the Soviets—and NASA. First the space agency had to choose a practical way of getting to the Moon and back.

Options

As had been the case when the U.S. was simply trying to put a man in space and get him back alive, scientists and engineers did not at first agree on the best way to get a man to the Moon and back. Three possibilities evolved: Direct Ascent, Earth Orbit Rendezvous, and Lunar Orbit Rendezvous.

Direct Ascent

Confronted with Kennedy's deadline, NASA at first planned to shoot straight for the Moon with a rocket that would blast off, descend to a lunar landing, and rocket back toward Earth, shedding stages all the way.

This was called the direct ascent method, but it had two major drawbacks:

◆ The booster required was huge, far bigger than anything in development

◆ The astronauts would have to land on the Moon atop a rocket stage hefty enough to lift them back to Earth

Imagine piloting an Atlas rocket back down to the launch pad while looking over your shoulder, and you get an idea of the engineers' misgivings.

Earth Orbit Rendezvous

An alternative was to launch two vehicles: a piloted spacecraft and a second, fully-fueled rocket booster. The manned spacecraft would go into orbit and then rendezvous and dock with the waiting lunar rocket, launched separately, and then use that second rocket to go to the Moon and return. This method was called Earth Orbit Rendezvous, and it had behind it the considerable clout of Wernher von Braun.

Lunar Orbit Rendezvous

But in 1962 a NASA engineer named John Houbolt came up with a novel approach that promised to save both time and money. He fleshed out earlier proposals to send one spacecraft into lunar orbit, and then detach from it a small landing craft that would set down on the lunar surface. After leaving the Moon, the landing craft would re-dock with the mother ship in lunar orbit. Only the mother ship would then return to Earth. This method was called the Lunar Orbit Rendezvous (LOR).

After wrestling with all the pros and cons, by mid-1962 even von Braun had been won over to LOR. The lunar orbit approach with its detachable landing craft was irresistible. Only the lander would have to descend to the Moon. Because of the weak lunar gravity, such a "lunar module" would not need much fuel to land on the Moon and get back into lunar orbit, so it could be made fairly small. This in turn meant that the entire Apollo spacecraft could be launched moonward by one Saturn V rocket, already in development by von Braun's team. Lunar Orbit Rendezvous became the key to meeting Kennedy's goal.

Apollo 1: Disaster at Pad 34

In early January 1967, NASA was ready to move on to Apollo from the successful Gemini program. First up was a test of the "mother ship," the Apollo command and service modules. On January 27, 1967, in preparation for Apollo-Saturn-204, the first piloted Apollo mission—and the first U.S. mission with a three-man crew—launch teams planned a simulated countdown that would start shortly before lift-off and work through the first few hours of the mission.

The astronauts—Virgil I. "Gus" Grissom, veteran of both the Mercury and Gemini programs; Edward White II, the first American to walk in space; and space rookie Roger B. Chaffee—would participate in the test at the Cape, while controllers at both KSC and Houston put them through their paces.

For this dress rehearsal, the astronauts were in their spacesuits, strapped in the Apollo capsule atop their Saturn IB rocket. There was no fuel in the booster's tanks, but to approximate orbital conditions as closely as possible, the astronauts were breathing pure oxygen. The men entered the capsule at 1:00 P.M.

The crew of Apollo 1, from left to right, Virgil I. "Gus" Grissom, Edward H. White II, and Roger B. Chaffee. They lost their lives in a January 27, 1967, fire in the Apollo command module during a test.

(NASA)

Astro Bio

Roger B. Chaffee   Lieutenant Commander Roger B. Chaffee was born on February 15, 1935, in Grand Rapids, Michigan. He received a B.S. in Aeronautical Engineering from Purdue University in 1957. He entered the U.S. Navy that same year and served as safety officer and quality control officer for Heavy Photographic Squadron 62 at the Naval Air Station in Jacksonville, Florida. In January 1963, he entered the Air Force Institute of Technology at Wright-Patterson Air Force Base, Ohio, to work on a Master of Science degree in Reliability Engineering. He logged more than 2,300 hours of flying time, including more than 2,000 hours in jet aircraft. In October 1963, Chaffee was chosen among the third group of astronauts selected by NASA. Along with the training program endured by all astronauts, Chaffee also worked with attitude and translation control systems, instrumentation systems, and communications systems for the Apollo program. On March 21, 1966, Chaffee was selected to go into space on the first Apollo mission, then designated AS-204.

A Foul Odor

The hatches were sealed with the men inside, and immediately there was a problem. Gus Grissom smelled a foul odor, like sour buttermilk. A small sample of his air supply was taken and, after some discussion, it was decided that the test should proceed.

Next an alarm went off indicating that the oxygen flow in the capsule was high. This alarm had gone off before and it was paid little heed. Once again, there was discussion and the test continued. Just as in space, the capsule was filled with pure oxygen, at a pressure about two psi higher than the sea-level pressure outside, which was 14.7 psi. A pure oxygen atmosphere was a known fire risk, but this test atop an unfueled rocket was considered a relatively safe operation. Engineers had neglected to consider the hazards of pure oxygen *combined with* the high pressure inside the cabin.

Space Talk

The **white room** is the small compartment on the gantry that fits around the spacecraft, providing a vestibule from which the astronauts enter the craft.

Communications Problems

More problems cropped up. The communications system between the spacecraft and the control room was not working. Later, there would be problems with communications on the launch pad. After some tinkering, launch control could hear all three astronauts, and so the test continued. The problem did, however, cause a 50-minute hold in the simulated countdown.

Faulty communications weren't unusual during spaceflight. As any observer of those early missions knew, the astronauts' transmissions were often difficult to understand. It didn't make any difference if they were 150 miles up in orbit or six feet away from the personnel in the *white room*. The delay to fix the communications system lasted from 5:40 P.M. until 6:31 P.M.

"Fire in the cockpit!"

At that time, ground instruments indicated another rise in the oxygen flow into the space-craft. Four seconds after that reading came the first horrible indication of the tragedy to follow. One of the astronauts said, "Fire. I smell fire."

Two seconds after that White could be heard, now more urgently: "Fire in the cockpit!"

Engineers and technicians who could see the inside of the spacecraft on television monitors watched in horror as visible flames licked upward and thickening smoke obscured their view.

There was no way to quickly get in or out of the spacecraft once the hatches had been sealed. The emergency hatch opening sequence was supposed to take 90 seconds, but the truth was that it had never been accomplished in trial runs at anywhere near that speed. From inside, a series of ratchet-type latches had to be undone to get out, and the first was behind Chaffee's chair in a spot where Grissom had to lower the chair's headrest just to get at the first one. White had made a portion of a full turn of the first ratchet when he was overcome with smoke.

Rescue Delayed

As the rescue crew in the white room outside the spacecraft tried to get in, the intense heat inside the spacecraft caused it to rupture, and the white room filled with heat and smoke. Five and a half minutes after the first alarm, the rescue team finally got the blistering hatch open. Doctors were three minutes behind the firemen, but there was little for them to do except officially pronounce the three men dead.

Although it is true that all three *Apollo 1* astronauts suffered major burns in the fire, those alone would not have been fatal. They had perished not from the fire, but from inhaling the toxic gases that filled the burning capsule. Death came from asphyxiation.

The loss of the crew of *Apollo 1* shook NASA to the core. Not until *Challenger* exploded nearly 20 years later, in 1986, would NASA see so dark a day. Meeting Kennedy's deadline now seemed doubtful.

Panel Investigates

Following the tragedy, a seven-member board, directed by NASA Langley Research Center Director Dr. Floyd L. Thompson, conducted a comprehensive investigation into the fatal fire. Their April 1967 report criticized the flammable nature of the atmosphere inside the spacecraft during the test and made specific recommendations leading to major design and engineering modifications. Test planning, test discipline, manufacturing processes and procedures, and quality control were all improved. The safety of the command and service module and the lunar module was increased.

The review board determined that conditions at the time of the fatal fire were "extremely hazardous" because of the pressurized oxygen atmosphere, the amount and location of combustibles in the command module, and the lack of a quick exit for the astronauts.

It is sadly ironic that one of the reasons *Apollo 1* had no quick-jettisoning escape hatch was because Gus Grissom himself had argued against it. Recall that it was Grissom whose Mercury capsule sank when his escape hatch accidentally blew (see Chapter 11, "Catching Up: Yuri Gagarin, Alan Shepard, and Project Mercury"). Grissom did not want that to happen again, and the removal of the explosive bolts on the hatch design for the Apollo spacecraft may have cost Grissom and his crewmates their lives.

Cosmic Facts

Regarding the risk he was taking as an astronaut, Gus Grissom once said, "If we die, we want people to accept it. We're in a risky business, and we hope that if anything happens to us it will not delay the program. The conquest of space is worth the risk of life."

Cosmic Facts

The fatal Apollo mission was originally known as AS (Apollo-Saturn)-204. Only after the tragedy was it redesignated *Apollo 1*.

The report also said that there should have been rescue personnel closer to the scene to react to the emergency. Fire and rescue personnel were not present in the "white room" outside the capsule when the fire started. The review board discovered that a full five minutes elapsed between the start of the fire and the time that rescue workers had all the capsule's hatches removed. The first firefighters did not arrive on the scene until eight to nine minutes after the fire started, and the first medical help did not arrive until 12 minutes after the fire broke out—though all agreed that there had been no chance to save the lives of any of the crew by the time the hatches were open. NASA had always flown astronauts in pure oxygen environments (it made the life support system lighter), but that practice was to change. Engineers replaced the complex and slow-to-open hatch with a new design that the astronauts could release and pop open in about ten seconds.

Apollo 7: First U.S. Three-Man Crew

There were no missions designated *Apollo 2* or *3*. *Apollo 4*, *5*, and *6* were unmanned missions to test the Saturn V rocket. And so it was *Apollo 7* that became the first manned Apollo mission. It was the only Apollo flight to be launched from Complex 34 at Cape Canaveral (and the scene of the *Apollo 1* fire). Launched atop a Saturn IB rocket, the mission got underway on October 11, 1968. All the rest of the Apollo moonflights required the larger Saturn V booster—however, the Skylab missions (see Chapter 23, "Skylab and the Space Detente") also flew Apollos atop a Saturn IB.

Dr. Jones's Corner

The Saturn IB, von Braun's forerunner to the Saturn V, was used on most manned Apollo missions which did not go to the Moon, but rather stayed in Earth orbit. A Saturn IB launched the first manned Apollo flight, *Apollo 7*, on October 11, 1968. Following the Apollo program, the Saturn IB was used three more times in 1973 to send men to the Skylab space station. It was next used in 1975, when it launched the U.S. crew for the Apollo-Soyuz Test Project. When used for an Apollo mission, the Saturn IB was 223 feet tall. Its first stage developed 1.6 million pounds of thrust at lift-off, using a cluster of fuel tanks taken directly from the old Redstone program.

The crew consisted of Commander Wally Schirra, who was a veteran of both the Mercury and Gemini programs, Command Module Pilot Donn F. Eisele, and Walter Cunningham as Lunar Module Pilot (though the mission carried no lunar module).

Astro Bio

Donn F. Eisele   Donn F. Eisele was born in Columbus, Ohio, on June 23, 1930. Following graduation from the U.S. Naval Academy, Eisele chose a career in the Air Force. He graduated from the Air Force Aerospace Research Pilot School at Edwards Air Force Base, California. He became a project engineer and experimental test pilot at the Air Force Special Weapons Center at Kirtland Air Force Base, New Mexico, and was one of the third group of astronauts selected by NASA in October 1963. In July 1972, Colonel Eisele retired from the Air Force and left the space program to become Director of the U.S. Peace Corps in Thailand. Upon returning from Thailand, Eisele went into private business.

Because of Tragedy, Changes

Twenty months had passed since the *Apollo 1* fire, and because of that tragedy, many changes had been made to the spacecraft. To alleviate the fire hazard prior to lift-off and during initial flight, the command module cabin atmosphere was composed of 40 percent nitrogen and only 60 percent oxygen. Not long after the astronauts reached space, the cabin atmosphere was gradually enriched to pure oxygen at a pressure of five pounds per square inch.

Astro Bio

Walter Cunningham   Walter Cunningham was born March 16, 1932, in Creston, Iowa. He earned a B.S. and an M.S., both in Physics, from UCLA. Cunningham joined the Navy in 1951 and began his flight training in 1952. In 1953, he joined a Marine squadron and served on active duty until August 1956. He worked as a scientist for the RAND Corporation prior to joining NASA. While with RAND, he worked on classified defense studies and problems of the Earth's magnetic field. Cunningham was one of the third group of astronauts selected by NASA in October 1963. He retired from NASA in 1971 and went into private business. Today he's a lecturer and a radio talk show host in Houston.

Also, for the first time, a fire extinguisher was part of the spacecraft's emergency supplies. Other new equipment on board included emergency oxygen masks and a TV camera, so the astronauts could broadcast television pictures live back to the Earth.

First Live TV

It is possible that NASA's public relations department went a little overboard during *Apollo* 7, and the result was the first near-mutiny in space. The mission flight plan included rehearsing just about every maneuver that would be necessary to land on the Moon, except for those involving the absent lunar module.

Dr. Jones's Corner

Schirra's mood aboard *Apollo 7* was not helped by the fact that shortly after launch he came down with a serious head cold. The virus soon spread to his crewmates. Soon everybody onboard had painfully congested sinuses, and their mood suffered accordingly.

Plus, this mission featured the first live television from space, and NASA wanted the folks at home to see as much of the mission as possible.

Commander Wally Schirra felt that the pressure of "performing" because of the TV schedule was making it more difficult than necessary for the crew to perform their duties. Sensitive to the pressure on his crew from an already packed flight plan, he got so fed up that at one point, Schirra refused to turn the camera on.

One of the primary functions of this mission was to test the service module's propulsion system. The big engine, essential for a return from lunar orbit, turned on and off when it was supposed to during both long and short burns. It would power the spacecraft into and out of lunar orbit, and correct the spacecraft's course as it headed toward the Moon.

After more than 10 days in space, *Apollo* 7 returned to Earth on October 22, 1968. The spacecraft splashed down in the Atlantic Ocean and was picked up by the USS *Essex*. Once again, Schirra had turned in a perfect engineering demonstration, despite his sour relations with Mission Control.

Apollo 8: Christmas at the Moon

NASA's next flight, *Apollo 8*, was to be a manned Earth-orbit test of the lunar module. But progress on the lunar lander was seriously behind schedule, and NASA, unwilling to merely repeat *Apollo 7*'s mission, took a chance on a very ambitious proposition. The Soviets had been testing an unmanned craft named Zond around the Moon, clearly preparing for a manned attempt to circle that body and steal the thunder of Apollo. Estimates were that the Russians would be ready sometime in the late fall of 1968. Faced with such a coup, NASA decided there was no reason to hold back. Without waiting for the lunar module, gutsy managers directed that *Apollo 8* ride its Saturn V to lunar orbit. It would be the first manned mission to leave Earth's orbit and go to the Moon.

Apollo 8's main objective was to test the command and service modules in lunar space. The mission lifted off from KSC on December 21, 1968, atop only the third Saturn V ever flown. The spacecraft did not carry a real lunar module, since there would be no lunar landing attempt. However, ballast in the Saturn V third stage simulated the weight of the real thing. This was the first time a Saturn V rocket had been used for a manned mission.

The rocket, with the Apollo spacecraft mounted on top, stood 363 feet tall, more than four times the height of the Redstone rocket that had pushed Alan Shepard into space eight years earlier. The first-stage Saturn V engines produced 7.5 million pounds of force at lift-off. (For more about the Saturn V rocket, see Chapter 19, "Apollo Hardware.")

The launch took place at Complex 39A at the Kennedy Space Center. This would be the starting point of all of NASA's manned lunar missions—except *Apollo 10*, which blasted off from nearby 39B. Together, the two pads would become known as the "Moonport."

Cosmic Facts

When the crew of *Apollo 8* blasted out of Earth orbit and set themselves on a course for the Moon, they were traveling faster than any human beings had ever moved before. The spacecraft accelerated to 24,226 miles per hour.

First Men to Leave Earth Orbit

Apollo 8's crew consisted of Col. Frank Borman, commander; Capt. James A. Lovell Jr., command module pilot; and Major William A. Anders, lunar module pilot.

Borman and Lovell had been in space together before, for 14 long days on *Gemini VII*. Anders was a space rookie. They were two hours and 50 minutes into their mission when they relit the Saturn V third stage and blasted out of Earth's orbit, headed for the Moon. This *trans-lunar injection* burn meant that the crew was now committed to deep space.

Space Talk

The firing of a spacecraft's engines to push it out of Earth orbit and onto a course for the Moon is called **trans-lunar injection**.

Cosmic Facts

During the 20 minutes that the spacecraft was on the far side of the Moon, during its first lunar orbit, there was no communication back to the Earth. There was a simple reason for this—radio waves travel in a straight line, and once the spacecraft swung behind the Moon, the Moon itself blocked transmissions.

Lunar Orbit

A little more than 69 hours after lift-off, *Apollo 8* slowed down by firing its service propulsion system engine and dropped into lunar orbit. The crew became the first human beings to see the far side of the Moon.

Back on Earth, Mission Control as well as the millions of people watching on TV waited anxiously for *Apollo 8* to emerge from behind the Moon. Houston's controllers cheered wildly when Lovell finally responded to their calls, "Go ahead, Houston."

Lovell then described for the world what he was seeing: "The Moon is essentially gray, no color. It looks like Plaster of Paris, like dirty beach sand with lots of footprints in it." The footprints were the thousands of craters pocking the lunar surface.

The astronauts made two elliptical orbits—168.5 by 60 nautical miles with an inclination of 12 degrees to the Equator—and then altered the shape of their orbit so that it was almost perfectly round, approximately 60 miles above the lunar surface. The crew was busy with photographic mapping, and describing the scene below for geologists back on Earth.

Prayers from Afar

The most memorable moments of the mission came on Christmas Eve, as the astronauts took turns reading from the Bible while a national audience watched live TV pictures being taken out the spacecraft's window.

Anders started by saying, "For all of the people on Earth, the crew of *Apollo 8* has a message we would like to send you. 'In the beginning, God created the Heaven and the earth ...'"

The Earthrise, as seen from Apollo 8 *while in lunar orbit.*

(NASA)

After taking turns reading, Borman concluded, "And from the crew of *Apollo 8*, we close with good night, good luck, a Merry Christmas, and God bless all of you—all of you on the good Earth."

I remember this incredible moment well. As a boy of 13, I watched, mesmerized, as the stark lunar surface rolled by on our TV screen. The beautiful Scriptural narration radioed down by the crew was the perfect accompaniment to our first glimpse of a newly revealed piece of Creation. Fittingly, the crew's photograph of a blue Earth rising above the cratered lunar horizon became one of the most famous images of the twentieth century.

After those eight circular orbits—for a grand total of 10 lunar orbits—the Apollo spacecraft again fired its service module engine, blasting out of lunar orbit and back toward Earth. The astronauts had been in orbit around the Moon for about 20 hours.

On December 27, after screaming back into the atmosphere at 25,000 miles per hour, *Apollo 8* splashed down in the Pacific Ocean, near the retrieval ship USS *Yorktown*. The mission was as flawless as any NASA had ever attempted. The goal of a lunar landing was coming tantalizingly close. But the Russians were not out of the game yet.

The Least You Need to Know

- Three astronauts—Gus Grissom, Edward White II, and Roger B. Chaffee—died on the launch pad when a fire broke out in the *Apollo 1* spacecraft during testing.
- The Apollo program was delayed over a year and a half by the *Apollo 1* tragedy.
- *Apollo 7* featured the first American three-man crew, and proved the redesigned spacecraft was ready for the Moon.
- *Apollo 8* was the first manned spacecraft to leave Earth orbit and go into orbit around the Moon.

Apollo Hardware

In This Chapter

- ◆ Saturn V
- ◆ Lunar module
- ◆ Docked modules
- ◆ Lunar orbit

The rocket that first took men to the Moon on *Apollo 8* was one of the engineering masterpieces of the twentieth century. To a large extent, the fulfillment of President Kennedy's goal—and the outcome of the race to the Moon—depended on the reliability of Wernher von Braun's giant: the Saturn V. Why was it needed? It took a lot more energy to propel three men to the Moon than it did to put two men into Earth orbit. And the spacecraft, including the new lunar module, was much bigger than either Mercury or Gemini. Because of these factors, NASA needed a bigger rocket than any of the military ICBMs. It came in the form of the Saturn V.

In this chapter, we'll learn about the rocket and spacecraft that the Apollo astronauts used to go to the Moon and about the first manned missions to visit our lunar neighbor.

The Saturn V Rocket

The Saturn V (pronounced "Saturn Five") was the last in a series of heavy-lift boosters that von Braun's team at the Army Ballistic Missile Agency began working on in the late 1950s. Tailored at first for lofting heavy military satellites, the Saturn series was taken over by NASA when von Braun's ABMA team became the nucleus of the Marshall Space Flight Center in Alabama. The Saturn boosters were NASA's answer to the robust Russian boosters of the early space race.

Research and development for the rocket had started way back in April of 1957 when Wernher von Braun ordered studies to be made of a four-engined booster that could achieve 7.5 million pounds of thrust.

At first von Braun called the new rocket the Super Jupiter, but the name was later changed to Saturn. There have been several Saturn models since then—the Saturn I and Saturn IB—with the Saturn V being the largest and most powerful rocket ever built.

The Saturn V rocket, if laid on its side, would be longer than a football field, including both end zones.

(NASA)

Dr. Jones's Corner

The Vehicle Assembly Building, where the Saturn V rocket was assembled, was a huge structure, 526 feet high, 513 feet wide, and 674 feet long. It was built with reinforced steel walls that could stand up to Cape Canaveral's hurricanes. Four fully stacked Saturn V's could fit inside the VAB's mammoth assembly bays. Each Saturn V was hauled to Launch Complex 39 on the back of a 3,000-ton crawler-transporter with eight tractor treads, each 40 feet long and 10 feet high. Its platform was the size of a baseball infield and it crept along with its precious cargo at a blistering one mile per hour.

All three stages of the Saturn V used liquid oxygen as the *oxidizer*. The first stage burned kerosene with the oxygen, while the fuel for the two upper stages was liquid hydrogen. Saturn V developed 7.75 million pounds of thrust at lift-off.

Space Talk

Remember, the **oxidizer** combines with the rocket fuel to support the rapid burning in the engine's combustion chamber. The Saturn V used liquid oxygen because it was denser than a gas; more oxygen could be pumped into the tanks than if it was stored in gaseous form.

First Flight of the Saturn V

The Marshall team under von Braun at first planned to test each Saturn V stage separately, working their way up to a flight of all three stages. This was the way they had succeeded with the Redstone, Jupiter, and earlier Saturns, gaining experience with each shot. But there wasn't time for this approach in the midst of a Moon race. NASA manager George Mueller ordered that the first Saturn V be launched "all-up," with all three stages live. The gamble was a tremendous success. The first Saturn V was launched into space in November of 1967, and it made a spectacular debut. On CBS-TV, Walter Cronkite reported the launch live and excitedly told America that "the whole building" was shaking. The crackling roar of the lift-off nearly shook the picture window behind him to pieces. The brilliant flames of the rocket could be seen 150 miles away.

Every stage worked perfectly on that first flight, but the early optimism that resulted did not last for long. The next test flight of the Saturn V came five months later, and this time an engine in the second stage shut down after about four and a half minutes. A second engine shut down soon thereafter. Later, the third stage refused to restart in orbit at all.

NASA scientists at first couldn't figure out what the problem was. The same rocket engines that were working during ground tests had failed in space. This turned out to be the clue that led to the solution.

The Saturn V was 363 feet tall. That's longer than a football field—including both end zones. The fully fueled booster weighed as much as a Navy destroyer— 6 million pounds. The 33-foot-wide first stage had five F-1 engines, each 19 feet tall. Each F-1 produced 1.5 million pounds of thrust at lift-off, for a total take-off thrust of more than 7.5 million pounds.

As it happened, the fuel lines that had broken from vibration in flight were holding while on the ground because of a reinforcing coat of ice that existed during ground tests. When stronger stainless steel pipes replaced the original corrugated lines, the vibration problem was solved. Getting the fix right the first time was essential. The budget and schedule did not allow for a third unmanned Saturn V test. The next time the rocket was fired, it would have to carry three astronauts.

Parts of the Saturn V Rocket

Let's take a look at "the stack," the Saturn V rocket with the Apollo spacecraft mounted atop it for a journey to the Moon. From the bottom up, the stack consisted of these six parts, rising like a totem pole from the launch platform:

- **The First Stage:** The huge S-IC powered the Saturn V up through the densest part of the atmosphere for the first two and a half minutes of the flight. During that time, its five F-1 engines gulped down 2,000 tons of kerosene and liquid oxygen. By the time the first stage jettisoned and the second stage roared to life, Saturn V was 41 miles above the Earth.

- **The Second Stage:** The S-II burned for six and a half minutes after the first stage had drained its tanks. Its five engines pushed the spacecraft to the brink of orbital speed before it ran out of fuel and gave way to the third stage. The S-II burned close to one ton of fuel per second.

- **The Third Stage/Lunar Module:** The third stage, or S-IVB, burned its single engine for two and a half minutes to place the spacecraft in a parking orbit around the Earth. Later, the S-IVB burned for another six minutes in order to push the spacecraft out of Earth orbit and to put it on a path to the Moon. The Lunar Module rode inside the top of the third stage, the S-IVB. With the third stage empty after trans-lunar injection, the Command Module turned around, docked with the Lunar Module, and pulled it out of the third stage.

- **The Service Module:** Attached beneath the command module, the service module carried breathing oxygen, control thrusters, the electricity-producing fuel cells, and propellants for the service propulsion system engine, needed to get into and out of lunar orbit.

- **The Command Module:** The command module was the cone-shaped space capsule in which the three astronauts lived during their trip to the Moon. It was the only part of the stack that returned to Earth.

◆ **The Launch Escape Tower:** This emergency rocket, like its Mercury ancestor, would have pulled the command module away from the rocket in case of a booster failure. The escape tower was jettisoned from the command module after second-stage ignition.

Docking with the LM

The lunar module (LM, pronounced "lem") as I've mentioned, rode inside the top half of the Saturn V's third stage. The third stage shut down about 12 minutes after lift-off, with both the spacecraft and attached third stage safely in Earth orbit.

But the third stage still had a job to do. It wasn't jettisoned, as most booster stages were after achieving orbit, but instead fired one more time, to get the spacecraft out of Earth orbit and on its way to the Moon.

With the three astronauts on their way to the Moon, the Command and Service Modules finally detached themselves from the third stage. The shroud that housed the lunar lander would fall away to expose the top of the LM. Then the crew turned the Command and Service Modules around using thruster rockets, so that they could dock with the LM. The narrow end of the cone-shaped Apollo capsule carried a probe that fit snugly into the LM's docking receptacle.

The Command and Service Module—usually known in shorthand as the CSM—would then back away from the S-IVB third stage, with the LM locked onto its nose. With the Apollo spacecraft now clear of the Saturn V's third stage, that booster was either sent into solar orbit, or deliberately crashed into the Moon to help reveal, through seismic waves, the structure of the Moon's interior.

The Apollo Spacecraft

As we have seen, the Apollo spacecraft consisted of the Command, Service, and Lunar Modules. While the service module provided power, oxygen, and rocket thrust, the Command Module was where the astronauts lived. The crew's food and equipment, as well as all the re-entry systems, were stored in the Command Module.

The Command Module was also equipped with a life-support system that kept the temperature comfortable, cleaned and recycled the air, and pressurized the cabin. Ensuring a constant pressure was important, because it allowed the astronauts to get out of their spacesuits except during critical phases of flight.

Both television and radio connected the astronauts to ground control. As was the case with the earlier manned space flights, the astronauts' vital signs were monitored from the Earth, using small medical sensors taped to their bodies.

Like Throwing a Pass

Firing a rocket at the Moon is a lot like a quarterback throwing a pass to a receiver who is running down the field. You don't aim where the receiver is, but at where he's going to be when the ball gets there.

In the same way, the Apollo capsule had to "lead" the Moon so that the two would converge in space and time. The flight path for the Apollo mission was figured out by a computer on the ground, but the crew also had the ability to compute their course independently. While the spacecraft was on its way to the Moon, slight alterations in its course were sometimes necessary, and were made using thrusters or the service module's big engine.

Computers at both Mission Control and the spacecraft were continually figuring and refining the positions, courses, and speeds of the Moon, the Earth, and the spacecraft. When mid-course corrections were called for, the computers would compute the time and size of the rocket pulse, and the astronauts would execute the burn to bring them back on target.

Cosmic Facts

Apollo 9's payload was the largest ever put into Earth orbit at that time. The Saturn V/Apollo 9 vehicle weighed 6,397,005 pounds at lift-off. Of that, 292,091 pounds made it into orbit, then a record.

Apollo 9: First Piloted Test of the Lunar Module

Apollo 9's crew consisted of Commander James A. McDivitt, veteran of *Gemini IV*; Com-mand Module Pilot David R. Scott, who had previously flown on *Gemini VIII*; and Lunar Module Pilot Russell L. Schweickart. Although *Apollo 9* did not go to the Moon as *Apollo 8* had, it took another important step toward landing a man on the Moon. *Apollo 9* was the first Apollo mission to carry all the machinery that would be needed to land on the Moon, including a lunar module, which was flown by astronauts for the first time.

Astro Bio

Rusty Schweickart   Russell L. "Rusty" Schweickart was born October 25, 1935, in Neptune, New Jersey. He attended MIT where he earned a B.S. and an M.S. in Aeronautics and Astronautics. Schweickart served as a pilot in the U.S. Air Force and Air National Guard from 1956 to 1963. Schweickart was a research scientist at MIT's Experimental Astronomy Laboratory and came to NASA as one of the 14 astronauts named in October 1963. He retired from NASA in July 1979 and went into public service.

Not only did *Apollo 9* take all that stuff into orbit, it checked it out and made sure it all worked together properly as well. The crew and ground team simulated all of the rendezvous and docking procedures that would be necessary to land on the Moon. The command-service module docked with the lunar module, and the astronauts successfully moved from one vehicle to the other and back again (through a small tunnel in the nose of the command module). The rendezvous and docking of the command-service modules with the lunar module took place six hours into the mission.

Another of the mission's highlights was a 37-minute EVA by astronaut Schweickart. Wearing the new "Moon-suit," the Apollo spacesuit meant for duty on the lunar surface, Schweickart stepped out on the porch of the lunar module and put the new suit through its paces. Later in the flight, the two spacecraft separated, leaving Scott aboard the command module while McDivitt and Schweickart maneuvered the LM freely in orbit. Their rendezvous with Scott had to succeed, because the spindly LM, built only to land on the Moon, could not make it down through the atmosphere alone. After notching one successful engine firing after another, and shedding the lower stage of the LM, McDivitt and Schweickart reunited their craft with Scott in the command module.

> **Cosmic Facts**
>
> In the "eagle eyes" department, astronauts of *Apollo 9* confirmed something that scientists had long suspected. It is easier to see in space. Because the distortions caused by the atmosphere are not present in space, the human eye can see things in space at distances that would be impossible on earth. For example, the *Apollo 9* astronauts saw the *Pegasus II* satellite from 1,000 miles away.

The Apollo 9 mission began on March 3, 1969, and lasted more than 10 days. The command module splashed down in the Atlantic Ocean on March 13 and was picked up by the USS *Guadalcanal*.

Down to the Last Nine Miles: *Apollo 10*

The Apollo program was clicking along like clockwork. When *Apollo 10* blasted off on May 18, 1969, it was the fourth manned mission NASA had launched in seven months. Once again, astronauts were going to the Moon, this time testing the lunar module while in lunar orbit. *Apollo 10* was the only Apollo mission to take off from Complex 39B at the Kennedy Space Center.

The *Apollo 10* crew consisted of Commander Thomas P. Stafford, who had been the pilot of *Gemini VI* and the commander of *Gemini IX*; Lunar Module Pilot Eugene Cernan, who had been the pilot on *Gemini IX*; and Command Module Pilot John W. Young, who had flown the corned beef sandwich with Gus Grissom on *Gemini 3* and had commanded *Gemini X*.

Of the three, both Young and Cernan would later walk on the Moon. Young would lead *Apollo 16*, while Cernan, as commander of *Apollo 17*, would become the last man in the twentieth century to walk on the Moon.

Simulation of the Landing to Come

All the events of the *Apollo 10 flight plan* were meant to rehearse those on *Apollo 11*, which promised to be the first attempted landing on the Moon. The only difference between the two missions was that *Apollo 10* was not to attempt to land. In fact, some NASA insiders wondered whether Stafford would attempt a surprise landing on *Apollo 10*, stealing the glory for his own crew. Stafford, though, was much too disciplined to try such a stunt, and the engineers later calculated that the *Apollo 10* LM was too heavy for a safe touchdown. *Apollo 11*'s shot at the history books was safe.

> **Space Talk**
>
> A **flight plan** is a detailed itinerary of events and tasks to execute during the course of a mission, designed to successfully accomplish the mission goals. Our thick flight plans—about an inch thick, typically—govern what the crew does every minute of every day. It's our roadmap for getting the mission done. On the space station, the flight plan for a given day is usually revised by Mission Control the night before and sent up to the crew's laptop computers for execution just before sleep begins.

The *Apollo 10* spacecraft had completed one and a half orbits of the Earth when the S-IVB booster stage was reignited. This sped the spacecraft up enough so that it could escape the pull of Earth's gravity and head out toward the Moon. After the three-day journey to the Moon, the spacecraft followed the same pattern as *Apollo 8* before it. It orbited the Moon twice in an elliptical orbit and then fired its rockets again to make its orbit circular, a constant 60 miles above the Moon's surface. Now the crew could focus on the landing dress rehearsal.

Making a Low Pass

On Day Five of the mission, Stafford and Cernan got into the lunar module, disconnected from the command-service module, and piloted the LM to within 47,000 feet of the Moon. Two passes were made over the eventual landing site for *Apollo 11*, and the astronauts reported it appeared smooth enough to try on *Apollo 11*.

Following this scouting duty, the LM dropped its landing, or "descent" stage, rendezvoused once again with the command-service module, and docked to rejoin John Young. Discarding their lunar module ascent stage, the crew prepared to leave lunar orbit.

Nearly six hours of live color television pictures were transmitted back to Earth, giving us our best view yet of the Moon's surface—as well as the spectacular view of the Earth as seen from the Moon.

Smooth Return

The service propulsion system was reignited one more time on May 24, thus blasting *Apollo 10* out of lunar orbit and setting it on a course for home. *Apollo 10* splashed down in the early afternoon of May 26, 1969, hitting its mark almost exactly, parachuting into the water less than four miles from its recovery ship.

NASA's days of teasing the Moon were now over. The next mission was the one with the simplest primary goal of all: To land a man on the Moon and return him safely to Earth—precisely the goal President Kennedy had set for the program eight years before. It was time to put the men and machines to their ultimate test.

The Least You Need to Know

- ◆ The Saturn V rocket was the most powerful ever built in its time (and still is in terms of weight hurled into orbit).
- ◆ The command module and lunar module docked shortly after Earth departure and travel linked together for the trip to the Moon.
- ◆ *Apollo 9* and *10* rehearsed all phases of the lunar landing missions to gain experience for the attempt to land on the Moon.
- ◆ *Apollo 10* did everything *but* land—Stafford and Cernan came within nine miles of the lunar surface.

Chapter 20

First Steps: Footprints for Eternity

In This Chapter

- ◆ Lift-off to destiny
- ◆ Landing of the *Eagle*
- ◆ *Apollo 12* struck by lightning
- ◆ Picking up our litter

It all came down to this—NASA's biggest moment. From the 15-minute flight of Alan Shepard to the first Earth-orbital flight of John Glenn to the orbiting of the Moon on Christmas, it had all been prelude. This was the main event. Humans were going to land on the Moon.

Right to the end, the Soviet Moon program was still lurking just offstage. The Russians had decided not to counter *Apollo 8*'s lunar orbit mission with a piloted flight to circle the Moon, but they were still hoping to steal the wind from Apollo's sails. In February 1969, the Soviets' N-1 Moon rocket failed disastrously in its first test flight. The last Soviet ace in the hole was a robot probe called Luna that might land ahead of *Apollo 11*, "scooping" the Americans by grabbing some soil from the Moon and returning it to Earth. A few days before *Apollo 11* launched, *Luna 15* headed for the Moon.

Neil Armstrong, the first man to walk on the Moon.

(Official NASA photograph)

Apollo 11: The Historic Voyage Begins

The crew was Commander Neil Armstrong, who had been the command pilot of *Gemini VIII*, Command Module Pilot Michael Collins, command pilot of *Gemini X*, and Lunar Module Pilot Edwin "Buzz" Aldrin, who was pilot of the last Gemini mission.

> **Space Talk**
>
> The **Apollo** space program was named after the Greek god Apollo. The name was the idea of Abe Silverstein, director of the Lewis Flight Propulsion Lab in Ohio. He later said, "I thought the image of the god Apollo riding his chariot across the Sun gave the best representation of the grand scale of the proposed program."

Aldrin recalls, "At breakfast early on the morning of the launch, Dr. Thomas Paine, the Administrator of NASA, told us that concern for our own safety must govern all our actions, and if anything looked wrong we were to abort the mission. He then made a most surprising and unprecedented statement: if we were forced to abort, we would be immediately recycled and assigned to the next landing attempt. What he said and how he said it was very reassuring." In other words, they shouldn't risk their lives because they feared this would be their only opportunity to walk on the Moon.

"Too Many Things That Can Go Wrong"

Collins remembers what he was thinking as he sat atop the Saturn V rocket:

> I am everlastingly thankful that I have flown before, and that this period of waiting atop a rocket is nothing new. I am just as tense this time, but the tenseness comes mostly from an appreciation of the enormity of our undertaking rather than from the unfamiliarity of the situation. I am far from certain that we will be able to fly the mission as planned. I think we will escape with our skins, or at least I will escape with mine, but I wouldn't give better than even odds on a successful landing and return. There are just too many things that can go wrong.

Apollo 11 blasted off from the Kennedy Space Center at 9:32 A.M., Eastern Daylight Time, on July 16, 1969. To those who watched the launch in person, the sound of the Saturn V was deafening, but inside the command module there was, according to Aldrin, "only a slight increase in the amount of background noise. In less than a minute we were traveling ahead of the speed of sound."

Before they returned, humanity's perception of its destiny and its universe would be forever altered.

Eleven minutes after lift-off, they were in Earth orbit. Three hours after lift-off, having left Earth orbit, the command and service modules (CSM)—known as *Columbia*—separated from the Saturn V rocket.

The combined command and service modules then turned around in space and docked with the lunar module (LM)—which was known as *Eagle*—pulling the lunar module out of its storage compartment in the third stage of the Saturn V. In that configuration, the craft hurtled toward the Moon. It would take three days to get there.

Anxious Moments Before Landing

Once in the vicinity of the Moon, the spacecraft fired its service module engine to put it into orbit around the Moon. When the craft was in lunar orbit, Armstrong and Aldrin clambered into *Eagle*.

Dr. Jones's Corner

A plaque was affixed to the front landing leg of *Eagle*. Signed by President Nixon and the crew, the plaque was etched with a map of the Earth, along with these words:

HERE MEN FROM THE PLANET EARTH

FIRST SET FOOT UPON THE MOON

JULY 1969 A.D.

WE CAME IN PEACE FOR ALL MANKIND

Buzz Aldrin, the second
man on the Moon.

(Official NASA photograph)

"You cats take it easy on the lunar surface," Collins said.

"OK, Mike," replied Aldrin with a grin.

With that, Collins flicked a switch that released the *Eagle*, which separated from *Columbia*. Armstrong, at the controls of the *Eagle*, performed a slow pirouette as he pulled away from *Columbia*.

"The *Eagle* has wings," Armstrong said joyously.

As the *Eagle* descended under rocket power toward the Moon, and was still 34,000 feet from the surface, the spacecraft's computer flashed a warning that it was overloaded.

The LM computer was choking on too much radar information flowing in from its sensors. This type of overload could overwhelm the computer, and in simulations was usually cause for an abort. Tense seconds went by as *Eagle* descended and Mission Control sized up the situation. Just in time, a sharp controller determined that the problem could be ignored—"We're GO on that alarm!" went up to the crew. The computer would keep complaining, but it would do its job through touchdown.

The third member of the
Apollo 11 *crew: Michael
Collins.*

(Official NASA photograph)

The Sea of Tranquility (the right eye of the "Man in the Moon," as he appears from Earth) was chosen for the first landing site. It was chosen because it is level and smooth—although it is, like most of the Moon, heavily scarred by meteor craters. About two minutes before the landing of the *Eagle*, commander Neil Armstrong had to take over from the automatic controls and fly past the originally chosen site; the guidance system had been taking him right into a sharp-rimmed crater nearly 600 feet in diameter, surrounded by boulders. Armstrong ended up landing the *Eagle* 1,300 feet west of the crater.

"The *Eagle* Has Landed"

As Armstrong drifted *Eagle* down toward the Moon's surface, the exhaust kicked up the finely grained soil that covered the Moon, creating a dust storm that inhibited Armstrong's visibility.

Later, Armstrong said, "This blowing dust became increasingly thicker. It was like landing in a fast-moving ground fog."

As the craft dropped the last few feet, Aldrin could be heard to say, "Drifting to the right a little. Contact light. Okay, engine stop."

Armstrong and Aldrin raise the American flag on the lunar surface.

(Official NASA photograph)

Mission Control radioed tensely, "We copy you down, *Eagle*." After a long pause, Armstrong replied with emotion, "Houston, Tranquility Base here, the *Eagle* has landed." The landing took place at 4:17:40 P.M., Eastern Daylight Time, on July 20, 1969.

Houston replied, "Roger, Tranquility, we copy you on the ground. You got a bunch of guys about to turn blue. We're breathing again. Thanks a lot."

Armstrong described the scenery:

Out of the window is a relatively level plain with a fairly large number of craters of the five- to 50-foot variety and some ridges, small, 20, 30 feet high, I would guess. We see some angular blocks out several hundred feet in front of us that are probably two feet in size. There is a hill in view.

The Moon race was over.

(Official NASA photograph)

"One Small Step ..."

The flight plan originally called for a sleep period right after landing, but no human beings could relax enough to sleep with their Moon walk beckoning. Over the next several hours, Armstrong and Aldrin prepared to leave the LM and explore the lunar surface. As you probably know, Armstrong went first. (Buzz had argued several months earlier that by Gemini EVA tradition, the LM pilot should be the first to step outside. Armstrong, the mission commander, overruled him.) He gave a running commentary as he went down the nine-foot ladder.

Cosmic Facts

Apollo 11 carried to the Moon and returned two large American flags, flags of the 50 states, District of Columbia and U.S. Territories, flags of other nations and that of the United Nations. The astronauts themselves were permitted only a few small personal mementos. No Moon rocks for souvenirs!

As Armstrong got to the bottom rung he said, "I'm at the foot of the ladder. Although the surface appears to be very, very fine-grained, as you get close to it, it's almost like a powder. I'm going to step off the *LM*."

Space Talk _____

LM is short for lunar module, the portion of the spacecraft that landed on the Moon. LM was usually pronounced *Lem*.

As Armstrong spoke these words, the world watched via a camera mounted on the outside of the LM. A little over a second after Armstrong set foot on the Moon, the world heard for the first time his now-famous words:

> That's one small step for man, one giant leap for Mankind.

Armstrong later said that, in his excitement, he had slightly altered what he had intended to say, claiming that the line was supposed to be, "That's one small step for a man … " But no one cared, and everyone got the point. No human could ever look at the Moon quite the same again.

The first step on the Moon took place at 10:56:15 P.M., Eastern Daylight Time on July 20, 1969. After a moment or two to catch his breath, Armstrong continued sharing his observations with the world: "The surface is fine and powdery. I can kick it up loosely with my toe."

Aldrin walks upon the "fine and powdery" lunar surface.

(Official NASA photograph)

Aldrin, once he joined Armstrong, proved to be less prolific but more poetic with his commentary, at one point describing his surroundings as a "magnificent desolation."

Aldrin's "magnificent desolation."

(Official NASA photograph)

The two astronauts discovered the "soil" at their landing site consisted of rock fragments ranging in size from fine, powdery particles to blocks of basalt three feet wide. While out on the surface, the crew set up a seismograph (to measure Moon quakes), a laser reflector (for measuring the Earth-Moon distance), and a solar wind collector.

Historic Phone Call

While on the Moon, the astronauts received a telephone call from President Richard Nixon. The president said, "Neil and Buzz, I am talking to you by telephone from the Oval Office at the White House, and this certainly has to be the most historic telephone call ever made …. Because of what you have done, the heavens have

Cosmic Facts

The first samples of Moon rocks returned to Earth were basalts, dark-colored igneous rocks. As lavas, they had erupted onto the Moon's surface about 3.7 billion years ago.

become a part of man's world. As you talk to us from the Sea of Tranquility, it inspires us to redouble our efforts to bring peace and tranquility to Earth ...."

Armstrong replied, "It's a great honor and privilege for us to be here, representing not only the United States but men of peace of all nations, and with interest and a curiosity and a vision for the future."

Astronauts Armstrong and Aldrin spent 2 hours and 31 minutes walking on the surface of the Moon. During that time they collected 47½ pounds of Moon rocks to bring back to Earth.

Equipment to measure Moon quakes, Earth-Moon distance, and to collect solar wind were set up by the crew of Apollo 11.

(Official NASA photograph)

Lift-Off from the Moon

Now it was time to head home. The *Eagle* was built in two parts. The descent stage and its fuel and engine got the astronauts safely to the surface. The ascent stage, with a separate engine, blasted off while using the descent stage beneath it as a launch pad.

Aldrin later wrote about the lunar lift-off: "The ascent stage of the LM separated, sending out a shower of brilliant insulation particles which had been ripped off from the thrust of the ascent engine. There was no time to sightsee. I was concentrating on the computers, and Neil was studying the attitude indicator."

So the bottom half of the lunar module remained on the Moon after the astronauts had left. (Today, there are six of those descent stages standing on the Moon as monuments to our urge to explore.) Lift-off from the Moon took place at 1:54:01 P.M., Eastern Daylight Time on July 21. The astronauts had spent 21 hours, 38 minutes, and 21 seconds on the Moon.

Dr. Jones's Corner

NASA scientists scratched their heads as they monitored the behavior of *Luna 15*. The Soviet probe went into orbit around the Moon before *Apollo 11*, and NASA asked the Soviets to keep clear of the latter's intended orbit. Instead of landing immediately, *Luna 15* continued to circle the Moon while *Eagle* landed, and Armstrong and Aldrin took their Moonwalk. Shortly afterward, *Luna 15* fired up its landing engines, dropped toward the Sea of Crises, and crashed into the lunar surface at over a hundred miles per hour. The Moon race was over.

Return Home

Rendezvous and docking in lunar orbit went smoothly. The lunar module docked with the command module. The astronauts returned, carrying their lunar samples with them, to the command module where they had a joyous reunion with Collins.

Collins recalls the moment that he was rejoined by his crewmates:

> The first one through is Buzz, with a big smile on his face. I grab his head, a hand on each temple, and am about to give him a smooch on the forehead, as a parent might greet an errant child; but then, embarrassed, I think better of it and grab his hand, and then Neil's. We cavort about a little bit, all smiles and giggles over our success, and then it's back to work as usual.

The LM was then jettisoned. The three men, using the service module engine, rocketed out of orbit with their priceless moon rocks. Before re-entry into the Earth's atmosphere, the command module separated from the service module.

The command module entered Earth's atmosphere at 25,000 miles per hour. After drifting downward beneath three parachutes, *Apollo 11* splashed down in the Pacific Ocean at 12:50:35 P.M., Eastern Daylight Time. The mission had lasted 195 hours, 18 minutes, 35 seconds from lift-off to splashdown. The retrieval ship was the U.S.S. *Hornet*. The astronauts had a strange homecoming different from that of any other astronauts.

Fear of Moon Germs

Scientists feared that there might be viruses or bacteria on the Moon that humans had no natural defense against. To avoid the risk of starting an epidemic, the *Apollo 11* astronauts were quarantined for three weeks after their return from the Moon, forced to talk to their families via telephone through a window, until it could be determined that they were germ free.

That same quarantine facility at Johnson Space Center is where shuttle astronauts undergo medical briefings both before and after launch. In fact, in a conference room you can still see that big picture window that separated the *Apollo 11* crew from their well-wishers.

Looking back more than 30 years after the first visit by humans to another world, the *Apollo 11* achievement has grown even more impressive. Using 1960s aerospace technology, and computers primitive compared to the plainest desktop PC of today, the United States sent three men to the Moon, and brought them back alive. All this was accomplished under the immense pressure of Russian competition and in full view of the rest of the world.

Chris Kraft, America's first Flight Director, was stunned when he first heard President Kennedy's speech committing America to a lunar landing: "… we've only put Shepard on a suborbital flight … men on the moon, has he lost his mind? Have I?" Given the U.S.A.'s 15 minutes of spaceflight experience in May 1961, Kennedy's bold decision to go to the Moon seems unbelievable in the risk-averse political climate of today. That single decision changed humanity's view of itself, made us a spacefaring species, and cast our future to horizons even more distant than the Moon. July 1969, and the twentieth century itself, will be remembered as a turning point in the history of civilization—the year humankind separated their destiny from that of their home planet, Earth.

Did We Really Land on the Moon?

There are a few skeptics out there (I believe unthinking ones) who believe the entire Apollo program was one big hoax. The confusion apparently began with a TV program called *Conspiracy Theory: Did We Land on the Moon?* The show put forth the theory that, to save money, the U.S. built a studio set of the Moon and used special effects to simulate the Moon landing.

According to this conspiracy theory, NASA made errors while faking the lunar landings, thus exposing the truth. The number-one error, it is claimed, is that in all the photos and films taken on the Moon, not a single star is visible. The true explanation for that is simple. It is impossible to photograph very bright and very dim objects at the same time.

With the camera set to photograph the sunlit Moon's surface and the men walking on it, the aperture was too small to allow in enough light to see stars. (Even your eyes can't see stars in space when the sun is out—your pupils get too small.)

Another supposed anomaly comes in the video of the first planting of the American flag on the Moon's surface. The flag appears to be rippling as if being affected by a breeze, yet there is no wind on the Moon. What people don't realize is how differently forces and reactions to forces behave on the Moon because the gravity is weaker. The astronauts twisted the flagpole back and forth in the soil to dig the pole in so that the flag would stand up on its own. As they did this the flag began to ripple. And, because there was less gravity and no atmosphere to stop the rippling, the flag continued to wave for far longer than it would have on Earth.

To further defend the authenticity of the Apollo program, NASA points out that the United States was not without enemies in 1969, including enemies who were sophisticated enough to monitor the radio broadcasts being transmitted by the astronauts, and make sure that they were coming from the direction of the Moon. They further point out the near half-ton of Moon rock brought back by various missions, samples of which have been studied all over the world. All those rocks are chemically discernible from Earth rocks, and therefore, rock-solid proof that we visited the lunar surface.

Department of Space Salvage: *Apollo 12*

After *Apollo 11*, what did NASA do for an encore? *Apollo 12*, of course—which would prove that the first lunar landing wasn't a fluke. The crew of Charles Conrad Jr., commander; Richard F. Gordon, command module pilot; and Alan L. Bean, lunar module pilot left the Earth at 11:22 in the morning on November 14, 1969. Along with the obvious, to land men on the Moon and return them to Earth, *Apollo 12* also had, as part of its mission objective, to retrieve parts of the unmanned *Surveyor 3*, which had landed on the Moon in April 1967.

Lightning Strikes

The mission got off to an exciting and dangerous start when the Saturn V booster rocket was struck by lightning—not once, but twice—soon after lift-off. The lightning strikes happened 32 seconds and 52 seconds after lift-off and caused a startling chain of alarms in the command module, temporarily cutting off electrical power and telemetry. The rugged Saturn V, though, rumbled skyward unperturbed. Quick action by Mission Control, and a thorough checkout in Earth orbit, saved an abort and got the mission back on track.

Clean-Up Crew

In a demonstration of pinpoint navigation and guidance, the astronauts found the *Surveyor 3*'s landing site, landed next to it, and set off on foot to examine that derelict spacecraft.

Conrad and Bean landed on the Moon on November 19 and stayed on the surface for more than 31½ hours. During their almost eight hours of walking on the lunar surface, the astronauts gathered approximately 76 pounds of Moon rocks.

Cosmic Facts

NASA was still analyzing the Moon rocks for any signs of harmful organisms or substances. So once again, the *Apollo 12* crew was forced to stay in quarantine for 21 days following their final Moon walk.

The astronauts even removed a TV camera and some other *Surveyor* parts, and returned them to Earth, the first job of cleaning up our litter on the Moon. Amazingly, scientists found that Earth bacteria had survived the harsh lunar conditions inside the *Surveyor* TV camera for nearly three years.

The *Apollo 12* astronauts left the Moon on November 20 and returned to Earth on November 24, 1969, splashing down in the Pacific Ocean where they were recovered by the USS *Hornet*.

Now NASA had run two consecutive successful missions to the Moon, and confidence was high. The next mission, however, *Apollo 13*, would remind us how risky flying to the Moon really was. *Apollo 13* would figuratively and literally bring everyone back down to Earth.

Astro Bio

Alan Bean   Alan Bean was born in 1932 in Wheeler, Texas. He began his Naval training in high school when at age 17 he joined the Naval Air Rescue service. Later, he was selected for a Navy ROTC Scholarship at the University of Texas, Austin. Upon graduation, he was awarded a B.S. in Aeronautical Engineering and was commissioned as an ensign in the United States Navy. Bean completed flight training in 1956 and was awarded his Naval Aviator Wings. After four years with Jet Attack Squadron 44 in Jacksonville, Florida, he was then selected for Naval Test Pilot School at Patuxent River, Maryland. Bean was chosen as an astronaut in 1963. He served in the backup flight crews for *Gemini X* and *Apollo 9* before becoming the lunar module pilot on the *Apollo 12* mission. In 1973, Bean returned to space as commander of the Skylab 3 mission, the second flight to the U.S.'s first space station. Throughout his career as a pilot and an astronaut, Bean's interest in art flourished. He began his formal artistic studies with night classes in drawing and watercolor while still a test pilot. During his years at NASA, he continued his art training in his free time, expanding his repertoire and developing a unique style of fantastic realism. In 1981, Bean resigned from NASA to devote time to painting and motivational public speaking. An accomplished explorer-artist, Bean creates paintings for future generations that help record humankind's first exploration of another world.

The Least You Need to Know

- The *Apollo 11* mission, with its crew of Neil Armstrong, Michael Collins, and Edwin Aldrin, was the first to land men on the Moon.

- NASA accomplished President Kennedy's lunar landing goal on July 20, 1969.

- Neil Armstrong, commander of *Apollo 11*, was the first man to walk on the Moon. Buzz Aldrin joined Armstrong on the Moon's surface a few minutes later.

- *Apollo 12* survived two lightning strikes soon after lift-off.

- *Apollo 12* astronauts executed a pinpoint landing next to the *Surveyor 3* robot lander on the Moon.

Chapter 21

Houston Solves a Problem

In This Chapter

- ◆ NASA thinks on its feet
- ◆ *Apollo 13* was both a failure and triumph
- ◆ "A pretty large bang"
- ◆ The world watches … and waits

NASA has had its share of human drama, whether it be the thrill of almost unthinkable accomplishment or the tragedy of lost life, but never before or since has NASA experienced such prolonged suspense and tension as it did during the *Apollo 13* mission.

Apollo 13: Courage and Ingenuity

Technically, *Apollo 13* was a failure. The mission was scheduled to land two men on the Moon, and it did not do that. But to call it a failure would be an insult to human courage and ingenuity.

There had been little thought of trouble. The lift-off from Kennedy Space Center at 2:12 in the afternoon of April 11, 1970, had gone well enough. Off to the Moon were James A. Lovell Jr., commander; John L. Swigert Jr., command module pilot; and Fred W. Haise Jr., lunar module pilot.

Astro Bios

Jack Swigert   John L. Swigert Jr. was born in Denver, Colorado, on August 30, 1931. He received a B.S. in Mechanical Engineering from the University of Colorado in 1953, an M.S. in Aerospace Science from the Rensselaer Polytechnic Institute in 1965, and an M.B.A. from the University of Hartford in 1967. Before joining NASA, Swigert held a position as engineering test pilot for North American Aviation, Inc. He was also an engineering test pilot for Pratt and Whitney from 1957 to 1964. He served with the Air Force from 1953 to 1956 and, upon graduation from the Pilot Training Program and Gunnery School at Nellis Air Force Base, Nevada, was assigned as a fighter pilot in Japan and Korea. After completing his tour of active duty with the Air Force, he served as a fighter pilot with the Massachusetts Air National Guard from September 1957 to March 1960 and as a member of the Connecticut Air National Guard from April 1960 to October 1965. He logged 7,200 flight hours. Swigert was one of the 19 astronauts selected by NASA in April 1966. He served as a member of the astronaut support crew for the *Apollo 7* mission. After *Apollo 13*, Swigert resigned from NASA and in August 1977 entered politics. In November 1982 he was elected by New Mexico to the U.S. House of Representatives. Tragically, he died of bone cancer on December 28, 1982, before he could be sworn in.

In fact, all had gone well for the first 56 hours of the mission, and the command and service module, attached of course to the lunar lander, was well on its way to the Moon when fate took control of the mission's fortunes.

"Houston, We've Had a Problem"

Mission Control had just had the astronauts take care of some routine checklist items, including some switch throws dealing with the fuel cells (electrical power generators) back in the service module. A few seconds later they heard a loud bang. The astronauts were not sure what the problem was right away, but it didn't sound, or look, good.

Cosmic Facts

Jack Swigert was originally assigned to the *Apollo 13* backup crew, and subsequently called upon to replace prime crewman Thomas K. Mattingly as command module pilot. The substitution was announced 72 hours prior to launch of the mission, following Mattingly's exposure to the German measles. The late replacement put Swigert in the middle of a life-and-death situation just a few days later. Ironically, Mattingly never came down with the measles.

The Explosion: A Play-by-Play Account

The *Apollo 13* astronauts were running through a routine checklist when they heard a loud bang. Here, verbatim, is the conversation aboard *Apollo 13* with Mission Control, from the explosion until five minutes after. The numbers at the left refer to hours, minutes and seconds since lift-off. The narrative includes significant events seen in spacecraft telemetry, and (in quotes) the crew and Mission Control calls.

55:55:20   Swigert: "Okay, Houston, we've had a problem here."

55:55:28   Duke: "This is Houston. Say again please."

55:55:35   Lovell: "Houston, we've had a problem. We've had a main B bus *undervolt*."

55:55:42   Duke: "Roger. Main B undervolt."

55:55:49   Oxygen tank No. 2 temperature begins steady drop lasting 59 seconds indicating a failed sensor.

> **Space Talk**
>
> **Undervolt** is short for undervoltage, meaning that they were receiving less electricity than anticipated due to the drop in output from the fuel cells.

55:56:10   Haise: "Okay. Right now, Houston, the voltage is—is looking good. And we had a pretty large bang associated with the caution and warning there. And as I recall, main B was the one that had an amp spike on it once before."

55:56:30   Duke: "Roger, Fred."

55:56:38   Oxygen tank No. 2 quantity becomes erratic for 69 seconds before assuming an off-scale low state, indicating a failed sensor.

55:56:54   Haise: "In the interim here, we're starting to go ahead and button up the tunnel again."

55:57:04   Haise: "That jolt must have rocked the sensor on—see now—oxygen quantity 2. It was oscillating down around 20 to 60 percent. Now it's full-scale high."

55:57:39   Master caution and warning triggered by DC main bus B undervoltage. Alarm is turned off in six seconds.

55:57:40   DC main bus B drops below 26.25 volts and continues to fall rapidly.

55:57:44   Lovell: "Okay. And we're looking at our service module RCS helium 1. We have—B is *barber pole* and D is barber pole, helium 2, D is barber pole, and secondary propellants, I have A and C barber pole." AC bus fails within two seconds.

55:57:45   Fuel cell 3 fails.

55:57:59   Fuel cell current begins to decrease.

Space Talk

Barber pole is the readout of an indicator on the instrument panel, intermediate between "open" and "closed," for example. In this transcript it means that the valve position of the RCS Helium tank went from "open" or "closed" on the indicator to a striped appearance, hence the barber pole call. This usually means the valve has lost power and the valve position cannot be determined. The big picture is that Lovell was saying he was getting a lot of weird instrument readings along with the bus losses.

55:58:02   Master caution and warning caused by AC bus 2 being reset.

55:58:06   Master caution and warning triggered by DC main bus undervoltage.

55:58:07   DC main bus A drops below 26.25 volts and in the next few seconds levels off at 25.5 volts.

55:58:07   Haise: "AC 2 is showing zip."

55:58:25   Haise: "Yes, we got a main bus A undervolt now, too, showing. It's reading about 25 and a half. Main B is reading zip right now."

56:00:06   Master caution and warning triggered by high hydrogen flow rate to fuel cell 2.

Short Circuit

A damaged circuit in oxygen tank #2 had started a fire in the tank's wiring insulation, leading to a violent explosion of the oxygen tank. That was the "bang" heard by the crew. The explosion also damaged oxygen tank #1 and cut off all oxygen supplies to the fuel cells. With the fuel cells dropping off line, a series of electrical failures soon confirmed that the spacecraft had lost the ability to generate electricity. The explosion had blown away panels on the outside of the spacecraft, which had in passing knocked the craft's main antenna out of alignment. The only power and oxygen left in the command module was the small quantity stored in batteries and tanks for re-entry.

Grappling with failing circuits and a spacecraft drifting out of control from the force of escaping oxygen, the astronauts knew nothing of what had caused the catastrophe. The oxygen vented quickly into space and then stopped when the tank went empty. So after some period of attitude disturbances, the ship got back under control of the thruster systems. The change in course to the moon was relatively minor; it was the disturbance to the pointing of the spacecraft that was critical for communications and guidance.

One thing they knew was they were in deep trouble. They were hundreds of thousands of miles from home, getting farther away with every second, and they had a command module with no power and oxygen. In Houston, NASA's best, the men who were trained to think on their feet, scrambled for a solution.

Making It Up as They Went

The controllers came up with a plan to power down the command module and move the men into the lunar module. Since the LM had its own batteries, its systems were still working.

The astronauts would use the lunar module—its power, oxygen, and engines—as their lifeboat. Designed to support two men for only a couple of days, it was doubtful its supplies could be stretched so that three men could make it back to Earth. The battle was to steer the spacecraft toward Earth and find a way to keep the crew alive until they could crawl back into the command module for re-entry.

Even with sufficient oxygen inside the LM, there was a dangerous buildup of carbon dioxide in a two-man spacecraft now supporting three. The command module had ample scrubber canisters, but its systems were dead. Therefore, the crew had to adapt the canisters to work in the LM. Mission Control and the astronauts worked together to jerry-rig an adapter so that a command module air scrubber canister could fit a LM hose. The crew used cardboard, plastic, and duct tape to link together components built by two different manufacturers.

Mission Control's hastily modified scrubber did the job—it chemically removed the CO_2 and averted a life-threatening crisis.

That problem solved, the LM oxygen supply looked like it might just make it to re-entry. Power would have to be conserved, and the temperature would drop dangerously low in the craft before they were through, but the new arrangements were survivable—barely. Now, how to get back home? Mission Control had one shot—target *Apollo 13* to loop around the back of the Moon, and use the Moon's gravity to whip it back

> ### Dr. Jones's Corner
>
> The *Apollo 13* astronauts found the cardboard they needed on the cover of their flight plan book. They got the plastic from storage bags they had on board. And duct tape? We still fly duct tape today on the space shuttle, three or four big rolls totaling hundreds of feet. Duct tape—don't leave home without it!

> ### Cosmic Facts
>
> The eyes of the world were on *Apollo 13*, from the moment the crisis occurred, until they were safely back on the Earth. Newspaper headlines all around the world, regardless of politics, tracked the story and asked for prayers for the crew of *Apollo 13*.

Cosmic Facts _____

As the men whipped behind the Moon—not in orbit, but rather being pulled around and sling-shotted back toward Earth—they were farther from Earth than any men had ever been, before or since. The crew of *Apollo 13* still holds the human record for distance from the home planet.

in the direction of the Earth. Without using the Moon's gravity, *Apollo 13* lacked the power to turn itself around and return home. Making up procedures from scratch, Houston directed the crew through two crucial engine firings, and got them on the way home. To further complicate matters, the astronauts couldn't navigate by eye. Their view of the stars was obstructed by debris hanging out of the hole in the service module.

Astro Bio

Fred W. Haise Jr.   Fred W. Haise Jr. was born November 14, 1933, in Biloxi, Mississippi. He received a B.S. with Honors in Aerospace Engineering from the University of Oklahoma in 1959, an honorary Doctorate of Science from Western Michigan University in 1970, and he graduated from the Air Force Aerospace Research Pilot School in 1964.

Haise was a Naval Aviation Cadet at Pensacola Naval Air Station, Florida, where he earned his wings. After his flight training, he served as a Tactics and All-Weather Flight Instructor in the Navy's Advanced Training Command at Kingsville Naval Air Station, Texas. From 1954 to 1956, Haise was a Marine Corps Fighter Pilot. From 1957 to 1959 he was a Fighter Interceptor Pilot in the Oklahoma Air National Guard, while a student at the University of Oklahoma.

After completing his degree, Haise worked for NASA for three years before being selected as an astronaut. During this period he worked as a civilian research pilot at the NASA Lewis Research Center in Cleveland, Ohio. Haise has logged 8,700 hours flying time. From 1963 to 1966, he was a Civilian Research Pilot at the NASA Flight Research Center at Edwards Air Force Base. Haise was one of 19 astronauts selected by NASA in April 1966. After his harrowing *Apollo 13* adventure and command of the first drop tests of the space shuttle *Enterprise*, Haise resigned from NASA on June 29, 1979 to join the Grumman Aerospace Corp. as Vice President for space programs. He is currently President of the Grumman Technical Services Inc. Division in Titusville, Florida.

The Slim-Fast Method

Dehydration was a problem. All three lost weight. The astronauts were forced to live on six ounces of water each per day. That's only 20 percent of the normal intake. That was supplemented to some degree by fruit juices. They ate hot dogs—when they ate.

In fact, the crew set an Apollo record for weight loss. They lost a total of 31½ pounds. Lovell lost 14 of those. These numbers are 50 percent higher than those of any other Apollo crew.

Sleep was impossible because of the cold. When the electrical system was turned off, the temperature plummeted to 38°F. The cold caused condensation to form on all the space-craft's inner surfaces, so it was both cold and wet. The astronauts later said that it was as if it were raining inside the spacecraft. The suffering was worst for Haise, who developed a kidney infection during the return home and spent much of the trip shivering with a fever. His crewmates say he never complained.

Incredibly, the makeshift plans and stop-gap solutions worked. Before re-entering the earth's atmosphere, the astronauts crawled back into their command module. Their lifeboat, the lunar module *Antares*, and the dead service module attached to the base of the *Apollo* capsule were jettisoned before the re-entry began.

The astronauts felt wistful as they watched *Antares* drift away from them. The ship had served them well, even if it had not gotten them to the Moon. They were stunned when they saw the service module after separation. There was a huge hole, 21 feet long and 6 feet wide, blasted in the side from the explosion, and wires and insulation hung in tatters near the missing panels.

With the entire world glued to television sets around the globe, the *Apollo 13* capsule made a blessedly normal re-entry.

After jettisoning the service module, the astronauts got to see, for the first time, the huge hole an explosion had blown into the side of their spacecraft.

(NASA)

Nixon Proclaims Success

President Richard Nixon, who had been scheduled to deliver a major speech that night on troop withdrawal from Vietnam, canceled the address and instead flew to the site of *Apollo 13*'s splashdown. He was aboard the USS *Iwo Jima* when the men were recovered from the sea. Weak with fatigue, dehydrated, and suffering from kidney infections, the crew of *Apollo 13* stood, amazingly, safely back on Earth.

Cosmic Facts

Jim Lovell's book on the ordeal of the *Apollo 13* astronauts was adapted into a much-acclaimed movie called *Apollo 13*, directed by Ron Howard, and starring Tom Hanks as Lovell.

In a speech televised around the world, Nixon officially proclaimed *Apollo 13* a success. Estimates show that the worldwide TV audience for the splashdown of *Apollo 13* may have been the largest in history.

The Least You Need to Know

- *Apollo 13* was to have made the third landing on the Moon in April 1970.
- An explosion in the service module on the way to the Moon crippled the *Apollo 13* spacecraft.
- NASA's Mission Control quickly improvised a plan to get the astronauts home safely.
- The endurance of the crew and the ingenuity of Mission Control turned *Apollo 13* into NASA's "successful failure."

Dune Buggies, Golf Clubs, and Especially Moon Rocks

In This Chapter

- ◆ Alan Shepard returns to space
- ◆ The lunar rover
- ◆ Exploring the mountains
- ◆ Last visit

Following the close call of *Apollo 13*, some called on NASA to slow the space program down a bit, or even stop going to the Moon altogether—but that isn't what happened. Instead NASA pursued its Moon exploration with an ever-more-ambitious series of what it called "advanced Moon missions."

Tee Time on the Moon: *Apollo 14*

Apollo 14 was headed for the Moon a year after *Apollo 13*, following an exhaustive review of Apollo systems and a redesign of the cryogenic storage tanks. The mission featured the return to space of Alan Shepard, America's first man in space (see Chapter 11, "Catching Up: Yuri Gagarin, Alan Shepard, and Project Mercury"). Long grounded by an inner-ear problem, Shepard's health had improved and he had once again been cleared for flight duty.

Astro Bio

Stuart Roosa   Stuart Allen Roosa (Colonel, USAF, Ret.) was born August 16, 1933, in Durango, Colorado. He earned a B.S. in Aeronautical Engineering from the University of Colorado. He served in the Air Force from 1953–76. He graduated from the Aerospace Research Pilots School and worked as an experimental test pilot at Edwards Air Force Base, California, from September 1965 to May 1966 before he became one of the 19 astronauts selected by NASA in April 1966. Following his role as crewmember on *Apollo 14*, he served as backup command module pilot for the *Apollo 16* and *17* missions, and was assigned to the space shuttle program until he retired in 1976. In the private sector, Roosa was a successful corporate executive. Roosa died on December 12, 1994, of a pancreas ailment.

Whereas Shepard's first mission in space had lasted about a quarter of an hour, this one would be considerably longer—and it would contain a perk that would make Shepard's long wait worthwhile: a stroll on the Moon.

Shepard was joined on the mission by a pair of space rookies: Stuart Roosa and Edgar Mitchell. During the 33 hours that Shepard and Mitchell spent on the Moon's surface, Roosa remained in lunar orbit aboard the command module, *Kittyhawk*, to observe and take photos of the Moon's geology.

Apollo 14 racked up a string of accomplishments and firsts. For example, up to that point, the mission was …

- ◆ The largest payload ever placed in lunar orbit.
- ◆ The largest payload returned from the lunar surface (more than 100 pounds of Moon rocks).
- ◆ The longest period spent on the Moon.
- ◆ The longest time spent walking on the Moon.
- ◆ The first use of color TV on the Moon.

Shepard and Mitchell landed at the Fra Mauro site that had been *Apollo 13*'s target. Once outside, they pulled along a small, wheeled tool cart to carry equipment, experiments, and their load of rock samples. Their investigations into lunar craters and geology were the most extensive yet.

But the simple truth is that this mission is always going to be remembered for one thing: It was on this mission that Alan Shepard smacked a golf ball with a seven iron, becoming the first human to play golf on the Moon.

Astro Bio

Edgar Mitchell   Edgar Dean Mitchell (Captain, U.S. Navy, Retired) was born in Hereford, Texas, on September 17, 1930, but considers Artesia, New Mexico, his hometown. He received a B.S. degree in Industrial Management from the Carnegie Institute of Technology in 1952, an M.S. degree in Aeronautical Engineering from the U.S. Naval Postgraduate School in 1961, and a Doctor of Science degree in Aeronautics/Astronautics from MIT in 1964. He entered the Navy in 1952. In May 1953, after completing instruction at the Officers' Candidate School at Newport, Rhode Island, Mitchell was commissioned as an ensign. He completed flight training in July 1954 and subsequently was assigned to Patrol Squadron 29 deployed to Okinawa. From 1957 to 1958, he flew A3 aircraft while assigned to Heavy Attack Squadron Two deployed aboard the USS *Bon Homme Richard* and USS *Ticonderoga*; and he was a research project pilot with Air Development Squadron Five until 1959.

Mitchell came to the Manned Spacecraft Center after graduating first in his class from the Air Force Aerospace Research Pilot School, where he was both student and instructor. Mitchell was in the group selected for astronaut training in April 1966. He served as a member of the astronaut support crew for *Apollo 9* and as backup lunar module pilot for *Apollo 10*. Following his time on the Moon with *Apollo 14*, he was designated to serve as backup lunar module pilot for *Apollo 16*. After retiring from NASA, he became Chairman of the Board, Forecast Systems Inc., Provo, Utah, and West Palm Beach, Florida.

Apollo 15: **The Happy Wanderers**

Apollo 15 left the Kennedy Space Center on schedule, at 9:34 A.M. EST on July 26, 1971. The primary objective was to survey and sample an intriguing geological region on the Moon—the Hadley-Apennine region, which included a canyon called Hadley Rille.

The *Apollo 15* command-service module also carried a new equipment bay that would enable astronauts to conduct geophysical and photographic studies of the lunar surface.

The crew consisted of David R. Scott, commander, veteran of both *Gemini VIII* and *Apollo 9*; Alfred Worden, command module pilot; and James B. Irwin, lunar module (LM) pilot. Both Worden and Irwin were making their first space voyages.

Dr. Jones's Corner

Sadly, the Russian space program experienced tragedy that summer. On June 16, 1971, the crew of *Soyuz 11*, who had been in space for 23 days aboard the new Salyut 1 Space Station, died during re-entry when a faulty spacecraft valve caused their cabin to lose pressurization. The three men had not been wearing space suits.

Astro Bio

Alfred Worden   Alfred Merrill Worden was born in Jackson, Michigan, on February 7, 1932. After graduating from the U.S. Military Academy in 1955, he earned an M.S. degree in Astronautical/Aeronautical Engineering and Instrumentation Engineering from the University of Michigan in 1963. After pilot training, he attended Randolph Air Force Base Instrument Pilots Instructor School in 1963, and served as a pilot and armament officer from March 1957 to May 1961 with the 95th Fighter Interceptor Squadron at Andrews Air Force Base, Maryland.

Worden was one of the 19 astronauts selected by NASA in April 1966. He served as a member of the astronaut support crew for the *Apollo 9* flight and as backup command module pilot for the *Apollo 12* flight. After Apollo, in 1972–1973 Worden was Senior Aerospace Scientist at the NASA Ames Research Center, and from 1973–1975, he was chief of the Systems Study Division at Ames. After retirement from active duty in 1975, Worden went into private business.

On July 30, Scott and Irwin in the LM fired its *descent propulsion system* engine. This landing phase, called "powered descent," took 12 minutes. If there had been difficulty landing, the LM came with enough fuel for 103 seconds of reserve hovering time. Astronauts Scott and Irwin stayed on the Moon for just shy of 67 hours. Once on the Moon, they deployed a scientific station and spent the rest of their 18 hours of Moon exploration in their brand new lunar rover.

The astronauts used their Moonmobile to drive to and collect samples from a long, narrow, winding valley—Hadley Rille—as well as from the bordering mountains called the Apennine highlands. Seventy-six kilograms (167½ pounds) of lunar material—soil and rock—were returned to the Earth.

Space Talk

The LM's **descent propulsion system** was the main engine of the descent stage, which lowered the lander to the Moon's surface. This descent engine could be throttled by the astronauts to enable the LM to hover.

The return to Earth was uneventful, right up until the deployment of the landing parachutes. Although all three main chutes deployed perfectly at first, one soon collapsed and the craft was forced to drift to splashdown on the two remaining chutes. Although it struck the water harder than usual, the impact did not injure the astronauts; they were safely recovered from the Pacific Ocean on August 7, 1971 by the USS *Okinawa*. The mission had lasted 295 hours and 11 minutes from lift-off to splashdown.

> **Dr. Jones's Corner**
>
> When the astronauts left the Moon, on August 2, they left the lunar rover behind. In fact, mounted on the rover was a television camera, and it was this camera that showed the pictures of *Apollo 15*'s lunar module blasting off from the Moon to once again dock with the command-service module in lunar orbit. That camera had been scheduled to work long after the LM was gone, and to provide images of an eclipse on August 6. Unfortunately, when turned on by Mission Control 40 hours after the astronauts' departure, the camera worked for only 13 minutes and then stopped.

First Use of the Lunar Rover

Apollo 15 featured the debut of the remarkable lunar rover, an automobile used by the astronauts to explore far more of the lunar terrain than would have been possible on foot. The rover, which had bucket seats and power steering, weighed 456 pounds and ran on two 36-volt batteries. It could travel 40 miles before the batteries were expended.

Stay Within Walking Distance

For safety reasons, NASA's flight rules stated that the astronauts were not to travel more than six miles from their LM. That way, in case of a rover breakdown, the astronauts would be able to walk back to their spacecraft before their life support backpacks were exhausted.

Assuming that the two astronauts and their life-support systems weighed 800 pounds (in Earth gravity), and that 238 pounds of photographic and communications equipment were onboard, the rover could carry a maximum of $59^{1}/_{2}$ pounds of lunar samples.

Cosmic Facts

In order to make it easier to transport aboard the lunar module, the rover folded up into a wedge less than half its operational size. The astronauts unpacked it from the lunar surface by deploying it from a compartment on the LM's descent stage.

Adapting to the Lunar Surface

The lunar rover was designed and built in just over a year and a half, by Boeing and the Delco Electronics Subdivision of General Motors. For NASA, 18 months between idea and execution was close to a speed record. The development of the lunar rover was not easy, however.

The rover had unique design features so that it could cope with the strange nature of the lunar soil, the extreme temperatures—260°F in the sun to –280° F in the shade—and the lower gravity. In order to get traction in the powdery soil of the Moon, the rover rolled on a set of specially designed wheels made of woven steel wire.

> ### Cosmic Facts
>
> Each Apollo mission to the Moon cost $400 million, a price tag that led some to wonder if the U.S. was getting its money's worth. For those who were not fascinated by field geology—that is, the collection of rocks—the repeated Apollo missions began to seem routine, and public interest slackened. Sensing this, and faced with the growing costs of former President Johnson's Great Society programs and the Vietnam War, Congress cut Apollo's budget, forcing the cancellation of the last three planned Apollo missions.

The wheels had chevron-shaped treads to dig into the lunar soil, and each wheel was given its own electric-drive motor. The driver could choose to steer with the front wheels, the back wheels, or with all four wheels. There was no steering wheel. All the controls (steering, acceleration, and braking) were on a joystick that was mounted between the astronauts, so that either man could drive.

> ### Cosmic Facts
>
> The lunar rover was definitely not ready for the Indy 500. It achieved an average speed of six miles per hour, and, when traveling on flatlands, moved as fast as 7.4 miles per hour. That was fast enough to kick up rooster tails of dust, flung from the rear wheels. At the end of the mission, the odometer on the rover revealed that it had been driven 17.3 miles on the Moon.

Navigation

Navigation presented another problem, since compasses don't work in the Moon's weak magnetic field, and taking a bee-line anywhere was impractical because of craters.

To help the astronauts determine where they were, and where they were going, a gyroscope and an odometer were connected to a tiny computer. The computer measured the distance and direction that had been traveled and thus determined the rover's location at any given moment. All done without GPS!

And You Thought Mexico City Was High!: *Apollo 16*

Apollo 16, the fifth landing of men on the Moon—and the second mission to include the use of the lunar rover—blasted off from the Kennedy Space Center on April 16, 1972. The *Apollo 16* lunar module set down on the lunar surface on April 21. The landing site was near the crater Descartes, the first landing in the Moon's highlands region. The elevation of that site was 7,400 feet higher than Neil Armstrong's landing site for *Apollo 11*.

Astro Bio

Charles Duke   Charles Moss Duke Jr. (Brigadier General, USAF, Ret.) was born in Charlotte, North Carolina, on October 3, 1935. Duke graduated from the U.S. Naval Academy in 1957 and earned an M.S. in Aeronautics from M.I.T in 1964. He graduated from the Aerospace Research Pilot School in September 1965 and stayed on there as an instructor. Duke was one of the 19 astronauts selected by NASA in April 1966. He served as member of the astronaut support crew for *Apollo 10*, as Capcom for *Apollo 11*, and as backup lunar module pilot for *Apollo 13*. In December 1975, Duke retired from the space program to enter private business.

Commander John Young and Lunar Module Pilot Charles M. Duke explored the lunar surface while Command Module Pilot T. K. Mattingly remained in lunar orbit. Using the lunar rover, the long-legged astronauts traversed 16.7 miles of lunar surface and collected more than 211 pounds of rocks during three sojourns outside totaling 20 hours and 14 minutes.

Astro Bio

Thomas Mattingly   Thomas K. Mattingly II (Rear Admiral, USN, Ret.) was born in Chicago, Illinois, March 17, 1936. He received a B.S. in Aeronautical Engineering from Auburn University in 1958. Mattingly began his Naval career as an ensign in 1958 and received his wings in 1960. He was then assigned to VA-35 and flew A1H aircraft aboard the USS *Saratoga* from 1960 to 1963. In July 1963, he served in VAH-11 aboard the USS *Franklin D. Roosevelt*, where he flew the A3B aircraft for two years. Mattingly was one of the 19 astronauts selected by NASA in April 1966. He served as a member of the astronaut support crews for the *Apollo 8* and *11* missions and was the astronaut representative in development and testing of the Apollo space suit and backpack (EMU). From December 1979 to April 1981, Mattingly headed the astronaut office ascent/entry group. He subsequently served as backup commander for STS-2 and STS-3, *Columbia*'s second and third orbital test flights. A veteran of three space flights, Mattingly has logged 504 hours in space, including 1 hour and 13 minutes of EVA during his *Apollo 16* flight. He was the command module pilot on *Apollo 16* (April 16–27, 1972) and the shuttle commander on STS-4 (June 26 to July 4, 1982) and STS 51-C (January 24–27, 1985). Mattingly resigned from NASA in 1985.

Both Young and Duke, and Mattingly orbiting overhead, took many different types of photos of the lunar surface. They used a variety of state-of-the-art equipment, including cameras that measured ultraviolet light.

Apollo 16 *astronaut John W. Young, while on the Moon, pulls tools from the aft end of the lunar rover.*

(NASA)

After 71 hours on the Moon, Young and Duke lifted off and rejoined Mattingly in the command module. They returned to Earth safely and splashed down in the Pacific on April 27, 1972, where they were picked up by the USS *Ticonderoga*. The biggest surprise of the mission was that the lunar samples the crew examined on the Moon, confirmed by post-flight study, were not the ancient lava flows scientists expected. Instead they were rocks made of pulverized fragments, fused together and tossed across the Moon as it was bombarded by asteroids and comets early in its history.

Good Night, Moon: *Apollo 17*

There are two things that most people remember about *Apollo 17*. The first is that it blasted off at night, creating a beautiful comet in the sky over Florida. The other is that it was the last manned mission to the Moon, and no one knows when we'll be going back.

In spite of the fact that the lift-off took place at 12:33 in the morning on December 7, 1972—or the "middle of the night," as many people refer to it—an estimated half-million people watched the lift-off in person, and, of course, millions more watched on TV. It was the first night launch of an Apollo-Saturn V.

At the moment of ignition, the flames beneath the Saturn V rocket became so bright that it was like daytime at the Cape. Because it was a cloudless night, the rocket could be seen as far as 300 miles away.

Aboard that rocket was more scientific equipment than had been carried by any previous Apollo mission. The astronauts aboard were Commander Eugene Cernan, veteran of *Gemini IX* and *Apollo 10*, Command Module Pilot Ronald E. Evans, and Lunar Module Pilot Harrison Schmitt.

First Scientist in Space

Apollo 17 was a landmark mission for another reason. It represented the first flight of a trained scientist to the Moon. Harrison "Jack" Schmitt, a geologist, was the first U.S. scientist in space.

Astro Bio
Harrison Schmitt   Harrison H. Schmitt, Ph.D., was born July 3, 1935, in Santa Rita, New Mexico. He received a B.S. in science from the California Institute of Technology in 1957; studied at the University of Oslo in Norway during 1957 and 1958; and received a doctorate in geology from Harvard University in 1964. Schmitt was a teaching fellow at Harvard in 1961 where he assisted in teaching a course in ore deposits. He did geological work for the Norwegian Geological Survey on the west coast of Norway and for the U.S. Geological Survey in New Mexico and Montana. He also worked for two summers as a geologist in southeastern Alaska. He has logged more than 2,100 hours flying time, including 1,600 hours in jet aircraft. Dr. Schmitt was selected as a scientist-astronaut by NASA in June 1965. In February 1974, Schmitt assumed additional duties as Chief of Scientist-Astronauts. In August of 1975, Dr. Schmitt resigned his post with NASA to run for the U.S. Senate in his home state of New Mexico. He was elected on November 2, 1976, with 57 percent of the votes cast. In January 1977, Schmitt began a six-year term as one of New Mexico's senators in Washington, D.C. He was defeated in his re-election bid and now is active as a consultant in both the government and science sectors. He is an outspoken advocate of returning science and commercial activity to the Moon.

Getting Beneath the Surface

For this final lunar mission, the command-service module was equipped with a special radar that could determine the physical properties of the lunar soil—and not just the surface, but beneath the Moon to a depth of almost a mile! This new information, when combined with the results of previous surveys, would complete a topographical profile of the Moon.

Astro Bio
Ronald E. Evans Ronald E. Evans was born November 10, 1933, in St. Francis, Kansas. He received a B.S. in Electrical Engineering from the University of Kansas in 1956 and an M.S. in Aeronautical Engineering from the U.S. Naval Postgraduate School in 1964. After receiving his commission as an ensign through the Navy ROTC Program at the University of Kansas, Evans completed flight training. In June 1957, Captain Evans was one of the 19 astronauts selected by NASA in April 1966. He retired from the U.S. Navy on April 30, 1976, with 21 years of service and remained active as an astronaut involved in the development of NASA's space shuttle program. He retired from NASA in March 1977 to become a coal industry executive. Evans died April 6, 1990, in Scottsdale, Arizona, of a heart attack.

Down in the Valley

Astronauts Cernan and Schmitt spent three days on the Moon, exploring an area known as the Taurus-Littrow Valley. They set up their experiments and explored in the lunar rover, conducting the most ambitious program of lunar exploration ever attempted. They deployed instruments, sampled the dark lavas of the valley floor, and knocked samples from boulders perched high on the surrounding mountains. They even found some orange soil, traces of a lava fountain that had erupted 3.7 billion years before. The odometer on the rover read close to 20 miles when they were through. *Apollo 17* brought back 242$\frac{1}{2}$ pounds of lunar material to Earth.

Cosmic Facts

Although the Saturn V rocket would be used again to loft the Skylab workshop into orbit, the roll-out to the launching pad of the Apollo-Saturn V rocket for the last manned mission to the Moon was treated as a solemn moment by those close to the space program. More than 5,000 spectators were said to be on hand as the huge rocket crept from the VAB and out the three miles to the launch pad.

Moon Bombs

The astronauts left behind equipment that would continue to monitor the Moon, including a seismograph. This was an "active" experiment, where geologists could set off explosive charges after the astronauts left, in the form of eight mortar bombs. The bombs were to be launched one at a time to cause an artificial "Moon quake," the vibrations of which would be measured by the instrument. Interpreted back on Earth, the Moon's shaking would give details about the structure of the Moon's crust below.

Other experiments left behind measured the effects of volcanic activity and meteor strikes. Even after the astronauts left the lunar surface and returned to the command-service module, they stayed in lunar orbit for another two days to complete their mapping mission around the Moon.

The astronauts departed the Moon on December 14, 1972, and returned to Earth on December 19. The crew splashed down in the Pacific and were picked up by the USS *Ticonderoga*. The entire mission, the last to the Moon, lasted just shy of 302 hours.

Space Talk

Also new among *Apollo 17*'s scientific equipment was a **spectrometer,** with which the astronauts could measure variations in the elemental makeup and density of the lunar atmosphere.

The Least You Need to Know

- ◆ Alan Shepard graphically demonstrated the weaker gravity of the Moon with a golf ball and a seven iron.
- ◆ The lunar rover was a specially built "moon jeep" that made exploration of the lunar surface easier and more wide-ranging.
- ◆ *Apollo 16* explored the Descartes Highlands, finding evidence of the early Moon's battered history.
- ◆ On the last Apollo mission, a scientist joined test pilots on space missions.
- ◆ The gathering and study of Moon rocks offered invaluable insights into the early history of the Moon, Earth, and solar system.

Part 5

After the Moon

In this part, I'll tell you about the Apollo-Soyuz Test Project, a joint orbital mission using both American and Russian piloted spacecraft. I'll also tell you more about my favorite space vehicle, the versatile and reusable space shuttle. After 20 years of service, it's still the world's safest and most capable spacecraft. You'll learn about the space shuttle's hardware, its early successes, and a tragic failure that could have been averted, the *Challenger* disaster.

To conclude, I'll take you on a behind-the-scenes tour of a space shuttle mission from the planning stages straight through to execution. I'll discuss NASA's cutting-edge research into our neighboring planets, our own Earth, and the outer reaches of the universe. Finally, we'll take a look at NASA's options for the future of human spaceflight.

Chapter 23

Skylab and Space Détente

In This Chapter

- ◆ Wounded space station
- ◆ Fixing on the fly
- ◆ Endurance records
- ◆ Apollo-Soyuz

The federal government's spending on the Vietnam War and growing social programs put the squeeze on NASA's budget even before the *Apollo 11* triumph. With the final three moon landings cancelled (you can see the Saturn V's for those missions at the Marshall, Kennedy, and Johnson Space Centers), NASA looked for its next act. Following the end of the Apollo program, NASA had a large amount of Apollo hardware left over, and so it chose to make use of it by going after a new goal—to establish a permanent human presence in space with an orbiting space station.

In addition, with Cold War tensions easing, plans were put in motion for the first time to work *with* the Soviet Union in some areas of space exploration and research, instead of racing against it. The two countries agreed to conduct a joint orbital mission by the mid-1970s.

A Look at Skylab

The first United States space station put into orbit was called Skylab 1. It was approximately 84 feet long and 22 feet diameter at its widest point. To save money and get the most out of the existing Apollo hardware, NASA's new Skylab workshop was converted from a Saturn V third stage, the S-IVB.

NASA stated that the purpose of the station was to analyze man's ability to work and live in space over extended periods of time. The crews were also assigned to conduct astronomical experiments and Earth studies from above the Earth's atmosphere. Space allowed astronomers a better view of the sun, stars, and planets because the view from Earth was distorted by the shifting layers of Earth's atmosphere. Above the atmosphere, free of dust, haze, and turbulence, the astronauts could make more precise astronomical observations.

The Skylab workshop was put into orbit first, on May 14, 1973, launched atop the last Saturn V rocket ever to fly. Three crews of three followed in late May, August, and November. The astronauts rode in Apollo command-service modules, launched atop Saturn IB rockets.

Damaged During Launch

Problems started right away for Skylab, with the initial launch of the space station itself. Vibrations during ascent caused a cylindrical shield designed to protect the station from meteorites to open up, and the slipstream—the air rushing by any rapidly moving object—ripped it right off, taking one of the station's two solar panels with it.

The meteorite shield did not rip away cleanly after it tore loose, and pieces of it wrapped around the remaining solar panel, keeping it from extending in orbit. With one panel gone and the other stuck closed, the new workshop was starved for electrical power.

Cosmic Facts

Skylab Crews

Skylab 2: Pete Conrad, Joe Kerwin, Paul Weitz

Skylab 3: Alan Bean, Jack Lousma, Owen Garriott

Skylab 4: Jerry Carr, Edward G. Gibson, Bill Pogue

With the loss of the meteoroid shield, the station's ability to control its own inner temperature was seriously affected. The temperature inside Skylab was rising to 126°F when it was in the sun. With equipment and supplies inside giving off gases due to the high internal temperatures, there was no guarantee that the air inside the space station was breathable.

Skylab 2

The original schedule called for the first
crew of three to go into orbit the day after
Skylab but, because of the problems with
the station's temperature, that initial manned
launch was postponed for 10 days. NASA scrambled to come up with a repair kit of tools and sun-
shades that might help them fix the damage.
Changes were made in the crew's space suits and
work schedule to make the extreme temperatures
less of a health hazard. The crew and Mission Control were gambling that humans in
space could assess the damage and come up with a solution that would salvage the space
station.

> **Cosmic Facts**
>
> *Skylab 2 Facts*
> Launch Date: May 25, 1973
> Days in Space: 28
> Orbits of the Earth: 404
> Splashdown: June 22, 1973,
> Pacific Ocean
> Recovery Ship: USS *Ticonderoga*

Trying to Fix the Damage

The Apollo craft was on its fifth orbit around the Earth when it rendezvoused and docked
with Skylab. Before docking, the crew gave the space station a once-over to visually analyze the damage that had been caused when the
station lost its micro-meteoroid shield.

They decided to attempt something that had
never been tried before. The astronauts were
going to try to fix Skylab right then and there
using a space walk. They depressurized the command module and opened the hatch (the one
redesigned after the *Apollo 1* fire). In a 40-minute
space walk, Paul Weitz first attempted to clear the
debris and free the jammed *solar array wing*.

> **Space Talk**
>
> Skylab's **solar array
> wing** was designed to gather
> up sunlight, and using solar or
> photovoltaic cells, convert it into
> electrical power for use by the
> station.

With his body extending out of the command module, and his legs being held by Joe
Kerwin as Pete Conrad piloted the ship, Weitz first tried to clear the debris with a 15-foot
pole ending in a shepherd's crook. This didn't work, so next he tried a prying tool. He
pushed against the stuck wing so hard that he actually shifted the CSM away from Skylab,
and Conrad had to fly some extremely close formation to hold him in there. Despite their
efforts, these first attempts didn't work, but everybody got an "A" for effort.

A Month in Space

The astronauts docked and finally boarded the workshop on their second day in space.
They had to endure some less than perfect conditions—high temperatures (up to 120°F)
and the threat of toxic gases. They entered tentatively, checking the quality of the air

before breathing it. The crew's first job was to extend a sunshield—looking like a parasol—out through an airlock opening, which did the job of lowering the temperatures to a manageable level.

Once they got moved in, these guests stayed, and stayed, and stayed. And they didn't leave until they had set the world record for longest time in space: 28 days. The endurance record broke that of the ill-fated crew of *Soyuz 11*, who in 1971 spent 23 days aboard the Russian space station but died during re-entry.

Astro Bio

Dr. Joseph Kerwin   Joseph P. Kerwin was born February 19, 1932, in Oak Park, Illinois. In 1953 he received a B.A. in Philosophy from the College of the Holy Cross in Worcester, Massachusetts, and an M.D. from Northwestern University Medical School, Chicago, Illinois, in 1957. Kerwin joined the Navy Medical Corps in 1958 and earned his wings at Beeville, Texas, in 1962. Kerwin was selected as a scientist-astronaut by NASA in June 1965. Following his trip into space Kerwin became the Director of Space and Life Sciences at the Johnson Space Center.

On the fourteenth day of the mission astronauts Conrad and Kerwin opened the Airlock Module hatch and ventured outside Skylab. This time, the astronauts succeeded in removing the debris and fully extending the jammed main solar array wing, restoring much of the electrical power to the space station. When the wing broke free, Joe Kerwin went flying, flung away from the workshop by the solar array, and reined in only by his tether. Conrad reeled him back in with a laugh. This EVA lasted 3 hours, 25 minutes.

Cosmic Facts

There was one other space walk during the mission, to change the film in the cameras mounted on the space station's solar telescope.

When it came time to go they left Skylab in much better condition than they had found her, floated back into their Apollo command module, and returned home, leaving the workshop to be occupied by the next crew.

Skylab 3: Docked for 59 Days

The crew of Skylab 2 did not hold the space endurance record for very long. In fact, the mark was broken by NASA's very next station crew: Skylab 3. With Commander Alan Bean, Pilot Jack Lousma, and Science Pilot Owen Garriott aboard, the crew was launched in their Apollo spacecraft on July 28, 1973. They didn't come down until almost two months later.

Astro Bio

Owen Garriott   Owen K. Garriott was born November 22, 1930, in Enid, Oklahoma. He received a B.S. in Electrical Engineering from the University of Oklahoma in 1953, and an M.S. and a Ph.D. in Electrical Engineering from Stanford University in 1957 and 1960, respectively. From 1961 until 1965, he taught Electronics, Electromagnetic Theory, and Ionospheric Physics as an associate professor in the Department of Electrical Engineering at Stanford University. Dr. Garriott was selected as a scientist-astronaut by NASA in June 1965. He then completed a 53-week course in flight training at Williams Air Force Base, Arizona. In addition to his space-flight on Skylab 3, he was a mission specialist aboard space shuttle mission STS-9. He resigned from NASA in 1986 and went into private business.

It looked at first as if those were going to be a very long two months, as all three crew members suffered through debilitating motion sickness during their first day aboard Skylab. There was a delay in turning on all the station's equipment until the green in the astronauts' faces faded.

A more serious problem emerged on the fifth day of the mission when there was an apparent failure of two of the command module's thrusters. If the thrusters were not working properly, they might have cut the mission short. In a worst-case scenario, it may have prevented the astronauts from getting back home safely.

Launch crews at the Kennedy Space Center worked overtime, getting ready in case a manned rescue mission needed to be sent into space. But the decision was made to continue with Skylab 3, and when it came time for the thrusters to perform, they did so perfectly.

There were three space walks on the mission. On August 6, Garriott and Lousma spent six and a half hours outside. During that time they were able to put a new and better sun shield on the space station, replacing the makeshift "parasol" shield that had been put in place by the crew of Skylab 2. They also replaced the film on the station's solar telescope cameras. The same two astronauts went for another EVA on August 24,

Cosmic Facts

Skylab 3 Facts
Launch Date: July 28, 1973
Days in Space: 59
Orbits of the earth: 858
Splashdown: September 25, 1973, Pacific Ocean
Recovery ship: USS *New Orleans*

Dr. Jones's Corner

The other notable function of this mission came under the education category, as the astronauts participated in classroom-oriented demonstrations regarding weightlessness. The roomy workshop, as big as a three-bedroom house, provided a terrific stage for the astronauts' televised performances in free-fall.

again changing the film and performing maintenance chores. The third space walk came on September 22, this time by Bean and Garriott, again changing film and completing minor repairs.

Skylab 4: All-Rookie Mission

The Skylab 4 mission got off to a rocky start. The mission was a few days from launch when cracks were found in the tail fins of the Saturn IB's first stage. The flight was delayed until the fins were repaired. Once over that hurdle, the launch, on November 16, 1973, and the rendezvous and docking with the space station went perfectly.

For the first time since the Gemini program, NASA launched an all-rookie crew into space. Commander Gerald P. Carr, Pilot William R. Pogue, and Science Pilot Edward G. Gibson all had an equal amount of spaceflight experience—zero.

The space endurance record held by the crew of Skylab 3 didn't last any longer than their predecessors' had. This rookie crew went from last to first in space time accrued, by virtue of their 84-day spaceflight. They traveled about 34.5 million miles during that time. At first the workload was heavy, and the astronauts, tired and frustrated, were still inefficient. In fact, the crew became so dissatisfied with the ever-mounting backlog of work that they declared an unscheduled day off to Mission Control and proceeded to turn off the radio while they got some rest. This mini-mutiny led to much-needed workload adjustments, and the crew soon hit their stride and exceeded their science objectives.

Astro Bio

Gerald P. Carr   Gerald Carr was born in Denver, Colorado, on August 22, 1932, but raised in Santa Ana, California. He received a Bachelor of Engineering degree in Mechanical Engineering from the University of Southern California in 1954, a B.S. in Aeronautical Engineering from the U.S. Naval Postgraduate School in 1961, and an M.S. in Aeronautical Engineering from Princeton in 1962. Upon graduation in 1954, he received his commission and subsequently reported to the U.S. Marine Corps Officers' Basic School at Quantico, Virginia. He received flight training at Pensacola, Florida, and Kingsville, Texas, and was then assigned to Marine All-Weather-Fighter-Squadron 114 where he gained experience in the F-9 and the F-6A Skyray. After postgraduate training, he served with Marine All-Weather-Fighter-Squadron 122, from 1962 to 1965, piloting the F-8 Crusader in the United States and the Far East. Colonel Carr was one of the 19 astronauts selected by NASA in April 1966. Carr retired from the United States Marine Corps in September of 1975 and from NASA in June of 1977.

The first of four space walks during the mission was conducted on November 22, 1973, and lasted 6 hours, 33 minutes. During the space walk, astronauts Pogue and Gibson took photos of Earth's atmosphere, installed experiments, and repaired an antenna housed on the space station's exterior. The second space walk, conducted by astronauts Carr and Pogue, occurred on Christmas Day, 1973. The astronauts changed film on the external telescope cameras. The space walk lasted seven hours, one minute. The third space walk took place on December 29, 1973. During this 3-hour, 29-minute space walk Carr and Gibson took advantage of their good timing and snapped extensive photographs of the comet Kohoutek, which happened to be passing through this portion of the solar system at that time. The fourth and final space walk of the mission was conducted on February 3, 1974, and lasted 5 hours, 19 minutes, again made to change film and retrieve experiments that gauged the effect of space exposure on various materials.

Skylab's End

NASA had planned that Skylab would remain aloft to await visits by the space shuttle, then in development. But the tight NASA budget and technical delays on the shuttle program, along with increased atmospheric drag, spelled doom for Skylab. With no way to boost its orbit to a higher altitude, Skylab re-entered Earth's atmosphere, burning up and dropping a few fragments across Australia on July 11, 1979. Leaving a fiery trail that made it resemble a shooting star, Skylab left behind an immense record of long-duration experience, and a legacy of 79 successful experiments.

Cosmic Facts

Skylab 4 Facts

Launch Date: November 16, 1973

Days in Space: 84

Orbits of the Earth: 1,214

Splashdown: February 8, 1974, Pacific Ocean

Recovery ship: USS *New Orleans*

The Apollo-Soyuz Test Project

The Space Race was officially over. We had won. Although the Cold War would not be officially over until the fall of communism in Russia, things had warmed up considerably since the Bay of Pigs/Cuban Missile Crisis days when the world teetered at the edge of nuclear war.

The time to compete against the Russians in space was over, and opportunities for cooperation were ripe. The first joint mission between the two countries—and the first piloted space mission to involve any two countries—was the Apollo-Soyuz Test Project (ASTP) in 1975.

Space Talk _____

The **docking module** was a special part of the Apollo vehicle put in place so that it could dock and function as one spacecraft with the Soviet Soyuz vehicle.

Cosmic Facts _____

After the two craft separated, Soyuz stayed in space for another two days before landing in the Soviet Union on July 21. The Apollo vehicle did not splash down near Hawaii until the twenty-fourth.

To pave the way for further joint missions, the ASTP was designed to test the rendezvous and docking compatibility of the U.S. Apollo and the Russian Soyuz spacecraft. To make docking possible—a difficult challenge because of different measurement units, design standards and pressurization systems—the U.S. built a *docking module* that would fit between the Apollo and Soyuz spacecraft. The Apollo spacecraft (officially *Apollo 18*) was basically the same as those that went to the Moon, but instead of extracting a lunar module from the Saturn IB upper stage, it docked with and pulled out the new docking module. At the front end was a docking collar that would work with the Russian system. *Soyuz 19* remained unchanged.

Soyuz went into space first on July 15, 1975. The Apollo launch came seven hours later. The two craft did not rendezvous and dock until July 17, a procedure that went smoothly. The astronauts (Vance Brand, Thomas Stafford, Donald "Deke" Slayton) and cosmonauts (Alexei Leonov, Valeri Kubasov) were able to crawl between the spacecraft and shake hands in space. They shared meals, mementos, and memories of this ground-breaking link-up in orbit.

Astro Bio

Deke Slayton   Donald K. "Deke" Slayton, the only member of the Original Seven not to man a Mercury flight, was born in Sparta, Wisconsin, on March 1, 1924. On his eighteenth birthday he enlisted as an aviation cadet in the Army Air Corps.

After receiving his wings in April 1943, he was sent to Europe where he flew 56 combat missions. Following the Allied victory over Germany, Deke was sent to the Pacific, where he flew seven more combat missions against Japan. After discharge from the Air Force in 1946, Slayton took twice the normal load of courses and earned his Bachelor's degree in Aeronautical Engineering in two years from the University of Minnesota.

After a stint working for Boeing as an engineer, he was recalled to active duty as a fighter pilot and maintenance officer at Bitburg, West Germany. He returned stateside in 1955 and completed the Air Force Test Pilot School at Edwards Air Force Base. Slayton had been a test pilot for more than three years when he became one of the Mercury 7. In March 1962, NASA announced that Slayton had a heart problem called idiopathic atrial fibrillation. It was a relatively minor condition, but NASA doctors grounded Deke.

continued

continued

Slayton remained busy behind the scenes, functioning as NASA's Coordinator of Astronaut Activities. When in July 1970 his heart problem disappeared, he returned to flight status. On February 9, 1973, NASA announced that Slayton would be part of the crew for the Apollo-Soyuz mission. On July 15, 1975, after 16 years of waiting, Slayton finally made it into space. He logged over 217 hours of flight time during his only spaceflight.

Slayton remained active at NASA until 1982. He died of brain cancer in June 1993.

Deke Gets His Chance

If the name Donald "Deke" Slayton rings a bell, it should. Deke was one of the original Mercury 7 astronauts, the only one who did not get to fly into space. He lost his flight assignment when a mild heart problem was detected.

Deke had stuck with NASA, and a physical eventually revealed no signs of the heart problem. Slayton was returned to flight duty and assigned to the crew of the ASTP. It had taken him 16 years of waiting, but he finally made it into space.

Astro Bio

Vance D. Brand   Vance DeVoe Brand was born in Longmont, Colorado, in May 9, 1931. He was a commissioned officer and naval aviator with the U.S. Marine Corps from 1953 to 1957. His assignments included a 15-month tour in Japan as a jet fighter pilot. Following release from active duty, Brand continued in Marine Corps Reserve and Air National Guard jet fighter squadrons until 1964. He received a B.S. in Business from the University of Colorado in 1953, a B.S. in Aeronautical Engineering from there in 1960, and an M.B.A. from UCLA in 1964. He was one of 19 astronauts selected by NASA in April 1966. Brand flew on four space missions: Apollo-Soyuz, and the space shuttle missions STS-5, STS 41-B, and STS-35.

Last U.S. Mission in Six Years

The working relationships established by NASA and its Soviet counterparts in getting the ASTP off the ground had a good effect on relations between the U.S. and U.S.S.R. The technical trust built up between the two space rivals helped ease tensions in other areas after Vietnam and the tense days of the Cold War.

What Skylab and the ASTP did not lead to, however, was an immediate encore. The U.S. had been to the Moon and had marked the symbolic end of the space race. What to do now? Because of budget difficulties and space policy differences, it took some time to answer that question. Six years would pass before NASA launched another manned spacecraft: the space shuttle.

The Least You Need to Know

- ◆ Skylab was the first U.S. space station and was occupied by three different crews.
- ◆ The space station was damaged during launch, but repaired by astronauts in space.
- ◆ The U.S. and U.S.S.R., after years of being competitors, worked together on the Apollo-Soyuz joint mission.
- ◆ Deke Slayton finally got his spaceflight reward on the Apollo-Soyuz Test Project mission.

Chapter 24

Launches Like a Rocket, Lands Like a Plane: The Space Shuttle

In This Chapter

◆ End of the disposables

◆ Anatomy of a shuttle

◆ Early tests

◆ A few firsts

NASA recognized that its post-Apollo budgets would not support a new, expensive effort to further explore the Moon, or to send astronauts to the surface of Mars. Despite calls for a Mars mission by Vice President Agnew, President Nixon was not interested in boosting NASA's funding in the face of his competing priorities, especially since the Moon race with the Soviets had been won.

Looking beyond Skylab, NASA proposed a winged space shuttle to enable cheaper flights to Skylab and future space stations. The thing that would make the space shuttle different from any other spacecraft ever built was that it would be reusable. The Mercury, Gemini, and Apollo missions had been made with spacecraft that were good for one-time use only. The parts that survived and came back might go into a museum, but they did not return to space. The space shuttle, on the other hand, would be used as often as needed.

Discovery is readied on Launch Pad 39A.

(NASA)

NASA hoped that the space shuttle, once built, would have lots of customers. By sizing it to carry military satellites, the agency hoped to grab that business from expendable rockets and make the shuttle the keystone of America's future launch systems. Cheaper, strategically vital, and a replacement for many existing rockets, the space shuttle was touted by NASA as the launch vehicle of the future. President Nixon approved the plan while John Young and Charlie Duke were on one of *Apollo 16*'s Moonwalks.

Besides reusability, the other characteristic that made the shuttle special was its unique landing capability. Although it launched like a rocket, straight up, propelled by two booster rockets and main engines fed from a huge fuel tank, it returned to Earth like an airplane and could land on any long runway. Initial NASA plans called for lowering launch costs by flying the shuttles into orbit nearly once a week.

How the Space Shuttle Works

A space shuttle consists of three main segments:

- A reusable delta-winged space plane, called an *orbiter*
- Two solid-propellant booster rockets, which are recovered and also reused
- An expendable tank containing liquid propellant for the orbiter's three main engines

The space shuttle takes off like a rocket.

(NASA)

An orbiter's three liquid fueled engines, drawing propellants from the external tank (or ET), and the two solid-propellant rocket boosters burn simultaneously for the first two minutes. Together, they produce about 7.3 million pounds of thrust at lift-off.

Following two minutes of flight, a space shuttle has reached an altitude of 32 miles by the time the boosters have burned all their propellant. At that point, they are jettisoned from the orbiter and ET and float down into the ocean on a parachute. They are then towed in by a ship and refurbished for use on a later shuttle mission.

The orbiter and ET continue toward Earth orbit, 100 miles up, using the liquid hydrogen and oxygen stored in the ET. When the orbiter's main engines cut off, just before achieving orbit, the external tank is jettisoned. It re-enters the atmosphere and breaks up, a few fragments falling into the Indian or Pacific Ocean.

Cosmic Facts

Only the orbiter portion of the space shuttle gets honored with a name. But it takes the entire stack of tank and boosters, not just the orbiter alone, to form the full space shuttle system.

Cosmic Facts

An assembled space shuttle (the orbiter, booster rockets, and ET) is approximately 184 feet long, 76 feet high to the tip of the orbiter's vertical tail, and 78 feet wide, measuring across the orbiter's wingtips. Lift-off weight is usually about 4,500,000 pounds.

The Vehicle Assembly Building (VAB) at the Kennedy Space Center, where the space shuttle orbiter is attached to its fuel tank and booster rockets before launch.

(NASA)

The orbiter continues to coast upward for a time, and then the onboard orbital maneuvering system (OMS) engines are fired. These place the vehicle in a near-circular low-Earth orbit. When the mission is over, the orbiter fires the OMS again to slow down about 50 minutes before landing. The shuttle free-falls from that point down to its encounter with the atmosphere at about 400,000 feet (around 75 statute miles high). From that point it's a 30-minute and 5,000 mile glide to landing at either the Kennedy Space Center or Edwards Air Force Base.

First Orbital Test: STS-1

The first four space shuttle launches were research and development missions, designed primarily to make sure that the system worked—and to fix the things that didn't.

Space Shuttle *Columbia* lifted off from Pad A on Launch Complex 39, Kennedy Space Center, on April 12, 1981. After a 54-hour, 36-orbit test-flight mission that verified the boosters, engines, and the ability of the orbiter to function in space, it landed on Runway

23, a dry lake bed at Edwards Air Force Base in California. The runway was so long that the orbiter overshot its intended landing mark by more than a mile and still made a safe landing. The crew consisted of Commander John W. Young—who was walking on the Moon when President Nixon approved the space shuttle program—and Pilot Robert Crippen. The STS-1 mission marked the first time that a new space vehicle carried a crew on its initial flight.

Orbiter Discovery *and its crew land on Kennedy Space Center's Shuttle Landing Facility Runway 15, creating a cloud of smoke as its wheels touch the concrete.*

(NASA)

Columbia STS-2

Seven months later, after a number of delays, *Columbia* went into space for the second time, marking the first time a spacecraft had been sent into space twice, on November 12, 1981.

Commander Joe Engle and Pilot Richard Truly stayed in orbit a little more than two days and ended with a landing on a dry Edwards Air Force Base lake bed. This time the landing was more accurate, with the orbiter overshooting its planned touchdown point by only 780 feet. The orbiter was then transported back east atop its Shuttle Carrier Aircraft.

Cosmic Facts

A special aircraft was developed to transport *Columbia* from California back to its home at the Kennedy Space Center. It was taken piggyback style, atop a specially modified Boeing 747 jumbo jet. This special airplane is referred to as a Shuttle Carrier Aircraft.

Discovery's *drag chute opens behind it as it slows on the KSC runway.*

(NASA)

Among the payloads taken into Earth orbit on mission STS-2 was something called the Office of Space and Terrestrial Applications experiments, better known as OSTA-1. OSTA-1 included a radar imaging processor that could, among other things, determine the location of natural resources on Earth. Known as the Shuttle Imaging Radar-A, the experiment revealed the existence of buried river channels beneath the sands of the Sahara. Also on this flight a Canadian-built robot arm known as the Remote Manipulator System was successfully operated for the first time. That arm would be used from then on to unload the payload from the shuttle's cargo bay. The STS-2 mission had been planned for five days, but the flight was cut short when one of the three fuel cells that produce electricity and drinking water failed.

Cosmic Facts

The space shuttle orbiter is attached to its two booster rockets and fuel tank inside the Kennedy Space Center's Vehicle Assembly Building (renamed from the Vertical Assembly Building after the Apollo moonlanding program ended). Stacked on its crawler-transporter, the shuttle, like the Saturn V, is then rolled out along a special road, called the crawlerway, to the launch pad. The crawlerway is as wide as an eight-lane highway, and a little bit more than three miles long. It's paved not with asphalt, but with pebbles of river rock and crushed gravel in order to absorb the weight of the massive caterpillar-like treads on the crawler.

Columbia STS-3: Testing of Alternate Landing Site

The third shuttle mission did not go as smoothly as the first two. Commander Jack Lousma and Pilot Gordon Fullerton launched aboard *Columbia* on March 22, 1982, and stayed in orbit for eight days. Amid a variety of scientific experiments, the crew exercised the remote manipulator arm and performed tests that measured the orbiter's ability to withstand the intense rays of the sun. The orbiter reoriented itself periodically so as to expose different portions of itself to the Sun. Yet another experiment measured the amount of contamination given off by the orbiter itself.

> **Dr. Jones's Corner**
>
> Today's operational fleet of four space shuttle orbiters are, in the order they were first flown: *Columbia* (1981), *Discovery* (1984), *Atlantis* (1985), and *Endeavour* (1992). *Challenger*, which first flew in 1983, was destroyed in a 1986 accident.

That was the good news. But the bad news started when both astronauts suffered from space sickness (vomiting), and the toilet began malfunctioning.

More problems came during the landing, which for the first time was attempted at the White Sands Missile Range in New Mexico (both the Cape and Edwards runways were socked in by bad weather). The abrasive gypsum dust at the White Sands runway contaminated the orbiter engines after landing, and extensive repairs were necessary before *Columbia* could fly again.

Columbia STS-4: A Job for the Pentagon

The research and development phase of the shuttle schedule ended with STS-4, which launched on June 27, 1982, with T. K. Mattingly as commander and Henry W. Hartsfield as pilot. The cargo consisted of nine scientific experiments provided by students from Utah State University, and a classified Air Force payload. The crew conducted a lightning survey with handheld cameras and performed medical experiments on themselves for two student projects. Landing occurred on July 4, 1982, on the 15,000-foot concrete Runway 22 at Edwards Air Force Base—the first shuttle landing on a concrete runway. President Reagan was on hand to congratulate the crew and declare the shuttle "operational." Few suspected that his confidence in the shuttle was premature.

Columbia STS-5: First U.S. Four-Person Crew

STS-5 was the first shuttle mission with a four-person crew. That crew consisted of Vance Brand, commander; Robert F. Overmyer, pilot; and the first mission specialists to fly the shuttle—Joseph P. Allen and William B. Lenoir. This was the first shuttle mission to launch satellites, in this case a pair of commercially owned communications satellites, which were put into orbit. A planned space walk—which would have been a first for a shuttle flight—had to be cancelled when problems with the space suits were discovered. I know exactly how Joe Allen and Bill Lenoir felt when they had to call off their space walk (see Chapter 1, "My Adventures in Space").

Challenger STS-6: First Flight of *Challenger*

For the sixth shuttle mission, *Columbia* got a break, and a new orbiter, *Challenger*, launched into space for the first time. STS-6 carried a crew of four: Paul J. Weitz, commander; Karol J. Bobko, pilot; and Donald H. Peterson and Story Musgrave, both mission specialists. Using new space suits designed specifically for the space shuttle, Peterson and Musgrave successfully accomplished the program's first space walk, performing various tests in the payload bay. Story, a scientist-astronaut from the Apollo days, had waited 16 years for his first trip into space.

The primary task of this mission was to deploy a $100 million tracking and data relay satellite. Although the two-and-a-half-ton satellite was successfully deployed from *Challenger*, its two-stage booster rocket shut down early, placing the satellite in a low elliptical orbit. Fortunately, the satellite contained extra fuel beyond what was needed for its planned orbital maneuvers. During the next several months the thrusters were fired at carefully planned intervals and gradually moved the satellite into its *geosynchronous orbit*. The $100-million satellite was saved.

Space Talk

A **geosynchronous orbit** is one in which the satellite stays over the same spot on the earth at all times. By putting a series of satellites in geosynchronous orbits around the Earth, it is possible to transmit a signal almost instantaneously around the world by bopping it from satellite to satellite in a geosynchronous relay. NASA planned to put a pair of such satellites into orbit to provide near-continuous communications with the shuttle.

Sally K. Ride: First American Woman in Space

When *Challenger* launched on STS-7 on June 18, 1983, the event received more press attention than any other flight since the debut of the space shuttle. For the first time, the U.S. was putting a woman astronaut into space: Dr. Sally K. Ride.

Astro Bio

Sally K. Ride Sally K. Ride, the first American woman in space, was born on May 26, 1951, in Los Angeles, California. She received a B.S. in Physics and a B.A. in English in 1973, and M.S. and Ph.D in Physics in 1975 and 1978, all from Stanford University. Dr. Ride was selected by NASA as an astronaut candidate in January 1978. She trained for a year and was made eligible to fly as a mission specialist on the space shuttle in August 1979. Before her own flight, she worked as a Capcom for the second and third space shuttle missions.

The rest of the crew on STS-7 consisted of Captain Robert L. Crippen (commander), Captain Frederick H. Hauck (pilot), and mission specialists Colonel John M. Fabian and Dr. Norman E. Thagard. This was the second flight for the orbiter *Challenger* and the first mission with a five-person crew.

Cosmic Facts

Tasks and experiments completed by STS-7: deployed satellites for Canada (ANIK C-2) and Indonesia (PALAPA B-1); operated the Canadian-built Remote Manipulator System (RMS) to perform the first deployment and retrieval exercise with the Shuttle Pallet Satellite (SPAS-01); conducted the first formation flying of the orbiter with a free-flying satellite (SPAS-01); carried and operated the first U.S.-German cooperative materials science payload (OSTA-2); operated the Continuous Flow Electrophoresis System (CFES); operated the Monodisperse Latex Reactor (MLR).

Dr. Ride's mission lasted 147 hours before landing on a lake bed runway at Edwards Air Force Base, California, on June 24, 1983. Sally had set the standard for women astronauts and became a role model for young women around the world.

Sally K. Ride, the first American woman in space.

(NASA)

Once again serving as a mission specialist, Dr. Ride launched with the largest crew to go into space to date. On October 5, 1984, she returned to space with six others on the mission designated STS-41-G. This time her crewmates were Captain Robert L. Crippen (spacecraft commander), Captain Jon A. McBride (pilot), mission specialists Dr. Kathryn D. Sullivan and Commander David C. Leestma, as well as two payload specialists,

Commander Marc Garneau and Mr. Paul Scully-Power. This flight stayed in orbit for eight days. During that time the crew deployed satellites, conducted radar observations of the earth, and during a space walk by Sullivan and Leestma, demonstrated how satellites could be "refueled."

The mission ended with a landing at Kennedy Space Center, Florida, on October 13, 1984. STS-41-G became the first mission to return from space and land at the new, three-mile-long runway at the Kennedy Space Center, carved out of the swamps near the VAB (Vehicle Assembly Building).

Dr. Ride was scheduled to go into space a third time, but canceled her training following the *Challenger* disaster (see Chapter 25, "Loss of Challenger") to serve as a member of a presidential commission designed to investigate the accident. Once that investigation was through, she stayed with NASA, working on long-range and strategic planning. Dr. Ride in 1989 joined the faculty at the University of California, San Diego, as a physics professor. She is also Director of the California Space Institute, a research institute of the University of California. She has written several books, including a children's book about her experiences called *To Space and Back*.

Guion S. Bluford Jr.: First African American in Space

Mission STS-8 saw *Challenger* back in space. For the first time, the shuttle launched at night, offering people all up and down the East Coast of the United States a view of its beautiful fireball as it rose toward orbit on August 30, 1983. Another first on this mission featured the first African American to fly in space, mission specialist Guion S. Bluford Jr. The rest of the five-member crew was composed of commander Richard H. Truly, making his second shuttle flight, Daniel C. Brandenstein, the pilot, and Bluford's fellow mission specialists Dale A. Gardner and William Thornton.

The night launch was necessary because of the tracking requirements of the primary payload, the Indian National Satellite, which was successfully deployed on the mission's second day. For use in various scientific experiments, human cells from the pancreas, kidney, and pituitary gland, as well as six live rats, were brought into space by STS-8. *Challenger*'s mission also became the first to pull off a night landing, touching down at Edwards Air Force Base on September 5, 1983.

Columbia STS-9: First Spacelab Mission

STS-9 saw the return of *Columbia* to active duty. Launched on November 28, 1983, STS-9 set a new record at the time; this mission included an all-time high six-member crew: John W. Young, commander, first astronaut to fly six times in space; Brewster H. Shaw, pilot; Owen Garriott and Robert A. Parker, both mission specialists; and Byron K.

Lichtenberg and Ulf Merbold, payload specialists. The latter two were the first non-career astronauts to fly on the Shuttle.

The STS-9 mission was devoted entirely to Spacelab 1, a joint NASA/European Space Agency (ESA) program designed to conduct advanced scientific research in space. The Spacelab module was carried in the rear of Columbia's cargo bay, connected to the crew cabin by a pressurized tunnel. Seventy-two scientific experiments were carried out in the fields of atmospheric and plasma physics, astronomy, solar physics, material sciences, technology, life sciences, and Earth observations. The effort went so well that the mission was extended an additional day to 10 days, making it the longest duration shuttle flight to date.

The space shuttle had been proclaimed operational. The research and development bugs had been worked out of the first reusable space vehicle—or so it was thought. But trouble, deadly trouble, lie ahead.

The Least You Need to Know

- The space shuttle is the most complex flying machine ever. It remains the world's most versatile and capable space vehicle.
- The shuttle is the first reusable space vehicle.
- The shuttle provided a new, reliable way to launch satellites and made an excellent science platform.
- Sally K. Ride was the first American woman in space, flying in 1983 aboard STS-7.
- Guion S. Bluford Jr. was the first African American in space, flying in 1983 aboard STS-8.

Loss of *Challenger*

In This Chapter

- ◆ "A major malfunction"
- ◆ A nation mourns
- ◆ Breakdown of a tragedy
- ◆ *Challenger*'s history

The sky was clear and the sun was bright on the morning of January 28, 1986, as the Kennedy Space Center prepared for the lift-off of Mission 51-L, the tenth flight of the orbiter *Challenger*, the twenty-fifth mission of the space shuttle. It was a beautiful morning, but very cold, and the spectators who had gathered in the area to watch the launch were appropriately bundled. In fact it was 36°F at Launch Pad 39B, a full 15 degrees colder than it had ever been before for a manned space launch. The launch had already been delayed five times because of the weather. Seventy-three seconds after lift-off, the launch vehicle exploded, and all onboard were lost.

Everybody remembers where they were on January 28, 1986, when they heard that *Challenger* had blown up. The tragedy stunned many Americans because it was totally unexpected. Fifteen years after *Apollo 13*, and 19 years and a day after the *Apollo 1* fire, we had come to take the safety of space travel for granted.

How It Happened

Pieces of the ruined space orbiter, its tank, and boosters rained into the Atlantic a few miles off shore. Even as a shocked nation watched the forlorn search for the crew's

Cosmic Facts

We do not know the precise origin of that first puff of smoke from alongside the *Challenger*'s right booster rocket, because the cameras set up to film close-ups of that area were, because of mechanical failure, not operating.

Cosmic Facts

The darkening color of the puffs of smoke indicated that grease and rubber were being burned away.

remains, NASA launched a sober investigation into the accident under direction of a presidential commission. A frame-by-frame examination of film taken of *Challenger*'s launch revealed symptoms of a problem less than three-quarters of a second after lift-off. A large puff of gray smoke squirted out of the side of *Challenger*'s right-side booster rocket.

By the time two and a half seconds had passed since lift-off, eight more puffs of smoke were seen, and the puffs were growing increasingly dark in color. At the 2.7-second mark one more puff of smoke came out, and then the problem stopped.

As we now know, those puffs were coming from a faulty joint on the side of *Challenger*'s right solid booster rocket, in an area that faces the craft's external fuel tank.

The Wind Shear

Thirty-seven seconds into the flight, the spacecraft encountered a serious *wind-shear* condition, the same condition that frequently is the cause of airplane turbulence, but *Challenger*'s guidance, navigation, and control system properly controlled the problem.

Cosmic Facts

Wind shear is a change in wind speed or direction, or both, over a short distance. Such changes help create eddies, or swirls of air, that cause turbulence. Wind shear can be vertical or horizontal and depending on the scale of shear can cause anything from minor turbulence to tornadoes.

Challenger cleared the wind shear at the 64-second mark. Just as on every shuttle launch, at about one minute into the flight, the shuttle's main engines and boosters reduced their thrust in a planned move to keep the air loads—pressures on the vehicle caused by its increasing speed and the density of the surrounding atmosphere—within limits. After passing through this region of greatest stress on the shuttle, the orbiter's computers throttled the main engines back up. The capcom in Mission Control, astronaut Dick Covey, gave the crew a call, "Challenger, GO at throttle up."

The frame-by-frame analysis shows that, just as the solid rocket boosters were increasing thrust, just before the 59-second mark, a very small flame appears in the right solid rocket booster, near the "aft field joint." That joint is a ring-shaped seal where two lower segments of the booster were joined at Kennedy Space Center.

Cosmic Facts _____

Adjusting to the sudden change in wind strength and direction caused by the wind shear required the largest course correction a space shuttle had ever executed this early in a mission.

A Steady Plume of Flame

By one minute after lift-off, the flame was a steady plume and it was playing on the outside of the external fuel tank and one of the struts that held the booster to the external tank. The appearance of the booster leak plume was a white flame. The exhaust from the main engines was a near-colorless blue. When the tank was breached, 64½ seconds into the flight, the extra hydrogen streamed into the main engine exhaust and colored it with a white streak.

Dr. Jones's Corner

Among those lost with *Challenger* was mission specialist Ronald E. McNair, 35, a physicist, amateur jazz saxophonist, and karate expert, who had been the second African American to travel in space. The *Challenger* mission was his second flight.

The weakened strut now gave way, allowing the solid rocket booster to swing into the ET, rupturing both fuel tanks and crashing into *Challenger*'s wing. Just before the 74-second mark, the stack veered out of control, and once hit sideways by the immense force of the supersonic windblast, the space shuttle tore apart. The ET fell apart, spilling its hydrogen and oxygen into a blazing white fireball. *Challenger* was traveling at Mach 1.92 (that is, 1.92 times the speed of sound) at an altitude of 46,000 feet. The orbiter and crew cabin were totally enveloped in the explosive burn. Large chunks of the orbiter sailed on upward, then fell free and plummeted into the Atlantic. There was no hope that the crew had escaped.

Aftermath

Immediately following the disaster, an ocean salvage operation commenced to recover as much of the orbiter as possible. The salvage operation lasted for seven months. During that time, the crew's remains, along with less than half of *Challenger*, were recovered from the sea.

The crew of Challenger, *lost in an explosion seconds after launch on January 28, 1986 (clockwise from left): Ronald E. McNair, Ellison S. Onizuka, Judith A. Resnick, Francis R. "Dick" Scobee, Michael J. Smith, Gregory Jarvis, and Sharon Christa McAuliffe.*

(NASA)

The presidential commission quickly zeroed in on the causes of the failure. Camera coverage revealed the plume of booster exhaust jetting from a gap in the right solid rocket booster (SRB) aft field joint. The gap had opened because of a design flaw in the joint seal. In the near-freezing temperatures on that January morning, the cold stiffened the rubber O-rings that filled gaps between the booster segment joint. When the metal joints flexed slightly open under the pressure of SRB ignition, the stiff O-rings failed to fill the gap, and white-hot rocket exhaust shot through the opening and eroded the steel casing. Under the flexing of wind shear forces, the joint reopened, and a blowtorch of exhaust gradually cut through the casing and the ET attachment strut. With this structural failure, the booster broke loose, ruptured the external tank, and dragged the shuttle stack out of control. The supersonic wind forces instantaneously tore the vehicle apart.

Cosmic Facts _____

Killed in the *Challenger* explosion were Francis Scobee, commander; Michael J. Smith, pilot; Judith A. Resnik, Ellison Onizuka, and Ronald E. McNair, mission specialists; S. Christa McAuliffe, teacher in space; and Gregory B. Jarvis, payload specialist. McAuliffe, 37—a Concord, New Hampshire, schoolteacher—had won a nationwide NASA competition to be the first teacher in space. After the tragedy, NASA restricted shuttle flights to professional astronauts. Previous "civilian" crew members included a senator, a congressman, and a Saudi prince. NASA is reevaluating its "civilians in space" policies now that orbital tourism is becoming a reality.

Worse than the design failure was the discovery that booster engineers, worried about the cold-weather vulnerability of their SRBs, had recommended canceling the launch that morning. But slow information flow and faulty decision-making by shuttle managers permitted the launch to go ahead, on the theory that no one had proven that it was *unsafe* to launch. NASA had gotten it backward—we shouldn't fly until we can prove that the risks are understood, and it *is* safe to launch.

The pieces of the spacecraft, tank, and boosters were analyzed to assist in the failure investigation. The wreckage was then transported to the Cape Canaveral Air Force Station. In January 1987 the pieces were placed in two abandoned Minuteman missile silos, and there they remain in long-term storage.

Dr. Jones's Corner

Challenger's commander, Dick Scobee, had a direct impact on my path to becoming an astronaut. I met Scobee and his wife, now June Scobee Rogers, at a 1985 talk he gave at the University of Arizona, where I was a graduate student. Since Scobee was also a U. of A. graduate, I asked him for advice on the right career path to pursue. He told me to just pick a scientific field I loved, and then do my best to excel in it. His encouragement, plus his advice to stick with it even in the face of failure, stood me in good stead in the years to follow. Five years after meeting him, I was offered the chance to join the ranks he had so ably served. June Scobee Rogers now heads the *Challenger* Educational Foundation, which continues the educational mission of her husband's crew through its "*Challenger* Centers" across the country.

History of *Challenger*

During the late 1970s, *Challenger* began its life as a test vehicle. It was constructed as part of the shuttle's development program, and during its early days, it underwent rigorous testing.

For an 11-month period, the craft underwent vibration testing to see if any frequency of vibration could cause *Challenger* to come apart at the seams—but the structure proved up to the task.

Space Talk

Challenger was named after an American Naval research vessel that sailed both the Atlantic and Pacific oceans during the 1870s.

Space Talk

Manned Maneuvering Units (MMUs) are jet packs that astronauts wore on their backs during several 1980s shuttle space walks. It enabled the astronauts to work free of their spacecraft during an EVA, so they could approach satellites and maneuver in space near the orbiter. The MMUs used compressed nitrogen gas jets to enable astronauts to move about and pivot in free-fall. A smaller version of the MMU, the SAFER jetpack, is worn as a safety device by EVA teams working on the space station.

Becoming Operational

Having fulfilled its testing role, *Challenger* was then moved to the Rockwell International plant in Palmdale, California, where it was converted from a test vehicle to a flight vehicle, thus becoming the second orbiter in NASA's fleet in July 1982.

Challenger made its first flight into space on April 4, 1983. It carried the first Tracking and Data Relay satellite into space on that maiden voyage.

Precedents

In missions that followed, *Challenger* became the first shuttle orbiter to be sent into space at night, a glorious sight for spectators throughout the Cape area. It was also a *Challenger* mission during which the first satellite capture and repair operation was successfully carried out. *Challenger* was the orbiter platform for three Spacelab missions.

Other firsts involved with *Challenger* missions include:

- First U.S. woman in space
- First African American in space
- First space walk by a U.S. woman
- First shuttle program space walk
- First untethered space walk (using *Manned Maneuvering Units*)

Workhorse

During the four years that *Challenger* was in operational use, it was the workhorse of the shuttle fleet, going into space more often and racking up more space time than any of the other orbiters.

Challenger spent a total of 69 days in space, circling the globe 987 times. A total of 60 men and women served aboard *Challenger* during its space missions.

Safety Is Always #1

The *Challenger* accident was indeed NASA's most costly failure, and the fatal launch could have been prevented with better decision-making and more effective communication. In the aftermath of the disaster, NASA rededicated itself to the principle that safety is always the primary concern. There have been 106 shuttle missions as of October 2001, and the loss of *Challenger* remains the only serious mishap.

Cosmic Facts

Challenger had come very, very close to disaster on a previous mission. It happened during August 1983, and again involved a problem with one of its solid rocket boosters. One booster's exhaust nozzle eroded excessively during its two-minute burn, and was eight seconds away from burning completely through—thus causing a catastrophic failure—when the rocket exhausted its propellant and stopped firing.

The *Challenger* disaster put U.S. human space travel on hold for nearly three years while NASA made a host of safety improvements and adopted design changes so that this accident could never happen again. The lessons of *Challenger* are still alive and part of the day-to-day safety culture at NASA.

The Least You Need to Know

- *Challenger* broke apart in flight on January 28, 1986, 73 seconds after lift-off.
- Seven people died in the explosion, including the first schoolteacher in space, S. Christa McAuliffe.
- The accident should have been prevented, but bad NASA communication and decision-making let *Challenger* lift-off with a fatal design flaw.
- *Challenger* had been the workhorse of the shuttle fleet at the time of the mishap.
- Sobered by the tragedy, NASA rededicated itself to safety, and its accident prevention record remains a good one.

Chapter 26

Return to Flight

In This Chapter

- New safety measures
- Geriatric journey
- Politicians and tourists
- Hubble, Hubble, toil and trouble

After *Challenger*'s loss, NASA made several important alterations to the space shuttle system. The solid rocket booster field joints were redesigned to provide a metal-to-metal seal between segments, a positive insulation seal on the inside of the joint, and an additional O-ring in the joint itself. Joint heaters were added to the booster casings to keep the O-rings at room temperature. Engineers made design changes that improved the safety of the shuttle's main engines. And the crew cabin received modifications that would enable the crew to bail out should the orbiter not be able to make it back to a runway.

Back in Business

By late summer of 1988, NASA was ready to test the waters of space again. *Discovery* lifted off from Pad 39B on September 29, 1988, with an all-veteran crew and a Tracking and Data Relay Satellite (TDRSS, pronounced Tee-Dress). The mission lasted four days and concluded with a landing at Edwards Air Force Base. The space transportation system was back in business.

But not in the commercial satellite business. President Reagan directed NASA to remove *commercial* payloads from the shuttle's manifest. Henceforth those satellites would fly on expendable boosters, and many large military payloads shifted back to the Air Force's Titan.

The shuttle's new role was to conduct scientific research, loft scientific satellites, and prepare for the construction of the new Space Station *Freedom*, approved by Reagan in 1984. For the next 10 years, the space shuttle would prove its worth as a versatile, reliable science platform in a series of more ambitious flights.

As for its space station role, construction was supposed to begin in the early 1990s. No one could have guessed how long it would be before the shuttle actually flew to that new destination.

Back to the Planets

One of the first tasks for the revived shuttle program was to clear the backlog of scientific spacecraft that had been stuck on the ground during the *Challenger* recovery. In May of 1989, Atlantis lifted off on STS-30 with the first of the new wave of planetary probes, the *Magellan* spacecraft, bound for Venus.

Magellan

Magellan was named after the first explorer to circumnavigate the world. It was launched on May 4, 1989, and its mission was to circle the globe of Venus and open it to thorough geological study.

Cosmic Facts

Taking pictures of Venus's surface had previously been impossible because of the planet's solid cloud cover, but *Magellan*, like Pioneer-Venus and several Russian craft, imaged the terrain using a synthetic aperture radar. *Magellan*'s instrument was much more powerful, though, uncovering surface features smaller than the size of a football field.

Launched from the shuttle on a solid-fueled Inertial Upper Stage (IUS) booster, the probe made one and a half loops around the Sun before going into orbit around Venus on August 10, 1990.

Ninety-eight percent of Venus's surface was mapped by *Magellan*, and precision radio tracking was used to measure Venus's gravity field and internal structure. In addition, the surface topography was mapped using radar altimetry.

Magellan, only 21 feet long by 30 feet across, made its final contribution to Venus research by measuring the density structure of the planet's atmosphere as it plummeted out of orbit, finally burning up on October 12, 1994.

Galileo

Next out of the gate was the *Galileo* Jupiter probe, launched aboard *Atlantis* on STS-34 in October 1989. *Galileo* was originally going to use the powerful, liquid-fueled Centaur upper stage, tossed out of the shuttle cargo bay. But that volatile mix of propellants was deemed too dangerous for the shuttle after *Challenger*, so *Galileo* was switched to the solid-fueled Inertial Upper Stage—and a slower trip to Jupiter.

Arriving in December 1995, *Galileo* sent an atmospheric probe screaming into Jupiter's cloud tops, where it radioed back composition, temperature, density, and pressure data until it was crushed by the overlying weight of atmosphere.

Cosmic Facts

We now know more about Venus' surface than about our own Earth's (where 70 percent of the surface is obscured by oceans).

Cosmic Facts

On February 3, 1994, Sergei Krikalev became the first cosmonaut to ride into space aboard an American spaceship—which, for the record, was the space shuttle *Discovery*.

Galileo continues to orbit Jupiter in late 2001, sending back thousands of images of the planet and its intriguing satellites, despite a jammed high-gain antenna that drastically cut the speed with which data could be sent Earthward. JPL engineers again demonstrated their ingenuity by reprogramming the craft to regain its scientific promise.

Hubble

One of the most anticipated space shuttle payloads after the return to flight was the Hubble Space Telescope, which carried a nearly eight-foot-diameter mirror designed to see to the edge of the universe.

Delayed four years by the *Challenger* accident, it promised to revolutionize astronomy by carrying its instruments above the obscuring haze of Earth's blanket of air.

Discovery lofted Hubble into orbit on STS-31 in April of 1990, gently deploying the telescope with the shuttle's robot arm. What followed was a scientific and public relations disaster for NASA.

Hubble's mirror had been ground mistakenly to the wrong curvature, and pre-flight testing had not caught the error. The telescope's vision was embarrassingly blurred. Not until December 1993, on STS-61, would Hubble recover its potential. The shuttle crew repaired the mirror with a series of spectacular space walks, and the telescope went on to provide us with views of the cosmos no one could have anticipated.

Upgraded periodically by shuttle crews with new instruments, Hubble promises to keep pushing back our frontiers of knowledge until at least 2010.

The Hubble Space Telescope can see and photograph things that aren't visible from Earth's surface.

(NASA)

Ulysses

The joint European-U.S. space probe *Ulysses* was next out of the shuttle's busy cargo bay, in October 1990 aboard STS-41 *Discovery*. Hurled by the IUS past Jupiter and over the poles of the sun, *Ulysses* explored regions of the solar wind and magnetic field that had never before been probed. It delivered a picture of the Sun's place in *interstellar* space that helps us understand how our star generates its prodigious energy and delivers it to the solar system.

Space Talk

Interstellar means located or taking place among the stars.

Dr. Jones's Corner

During my tenure in the Astronaut Office, the shuttle was engaged in three main enterprises:

◆ First, it launched scientific probes into Earth orbit and across the solar system.

◆ Second, it served as a scientific research platform for a series of Spacelab missions and for space- and Earth-looking instruments like the Astro telescopes and the Space Radar Lab advanced imaging system (I helped operate the latter system aboard *Endeavour* in 1994).

◆ Third, the shuttle began preparing for space station missions by flying to the Russian Mir station, and in the end, lofting the first American elements of the new International Space Station. But that's a story for Chapter 27, "The International Space Station—Call Sign 'Alpha'."

Two "Great Observatories"

With the Hubble Space Telescope and the Compton Gamma-Ray Observatory, launched aboard STS-37 in April of 1991, NASA had two "Great Observatories" in orbit and examining the cosmos. The third in the series was the Advanced X-Ray Astronomy Facility (AXAF), renamed after launch the Chandra X-Ray Observatory.

Measuring Gamma Rays

The Compton Gamma Ray Observatory, placed into orbit by Atlantis on April 5, 1991, was a space-based observatory that studied the universe by making a systematic survey of the natural sources of gamma rays.

At 35,000 pounds, it was the heaviest satellite ever put into orbit by the space shuttle. The CGRO mission lasted nine years. It ended on June 4, 2000, when NASA redirected the craft back into the Earth's atmosphere.

Chandra

Chandra roared into orbit aboard the shuttle on STS-93 in July 1999, and its IUS upper stage took it into an elliptical orbit soaring over 100,000 miles from Earth. The telescope can "see" x-rays with a resolution equal to being able to read a one-centimeter-high letter from a distance of slightly less than half a kilometer.

Cosmic Facts

Read about the latest Chandra's discoveries by checking out chandra. nasa.gov and chandra. harvard.edu.

Since launch, Chandra's been looking at the most violent processes in the universe, collecting the x-rays streaming from exotic celestial objects, matter falling into black holes, and stellar explosions. Chandra has equaled in the x-ray universe the revelations made by the Hubble Telescope in the visual part of the spectrum.

John Glenn's Return to Space

The Old Man and the New Sea Department: NASA Administrator Dan Goldin made an announcement on January 16, 1998, that was to excite public interest in the space program to a degree unknown since Americans had walked on the Moon.

Goldin announced that space shuttle mission STS-95 would go into orbit on October 29 of that year, and that among the crew for the mission would be former astronaut and senator from Ohio, John Glenn, the first American to go into orbit 36 years before.

At 77 years, Glenn would be the oldest man ever to go into space, which would give researchers an opportunity to examine the effects of space travel on older people. Many of the effects of free-fall on astronauts echo the characteristics of aging, and NASA thought experiments on Glenn could help explore this parallel. So Glenn was used as a human guinea pig, though he earned the title payload specialist. NASA teamed up for the experiments on Glenn in conjunction with the National Institute on Aging.

One experiment Glenn operated studied the way certain proteins are processed during weightlessness. Muscles tend to weaken after a prolonged period in space, and scientists were attempting to find out how that process occurs.

In another investigation, Glenn's alertness at different hours was measured—as was that of the rest of the crew—to determine if his sleep patterns in space developed differently. Glenn was also the flight's photography specialist.

Investigators are waiting for the next research mission (STS-107 in the summer of 2002) so they can compare Glenn's data to the astronaut subjects on that flight. The scientists want more than just a couple of test subjects before they draw conclusions. So Glenn's experiments await the context of other investigations to be flown soon on the shuttle and space station.

STS-95 was noteworthy in other ways. It carried the Space Hab research module back behind the crew cabin and deployed and retrieved the Spartan solar-observing spacecraft. *Discovery* once again demonstrated the scientific versatility of the space shuttle system. (Space Hab is short for Space Habitat, by the way. It's a private company that builds modules for the space station.)

John Glenn, the first American in orbit, returned to space in 1998, 36 years after his historic first flight.

(NASA)

Politicians in Space

John Glenn served as a payload specialist on STS-95, a job description usually applied to technical experts who train for a specific research job on a single flight. Their expertise on that particular payload is so valuable that it wouldn't be practical to train a mission specialist up to that same standard.

Payload specialists on the shuttle have included physicists, astronomers, oceanographers, defense specialists, biomedical researchers, and even veterinarians. But perhaps the most unusual in the menagerie of payload specialists were two who were, like Glenn, politicians.

In the year before *Challenger*'s loss, NASA flew two members of Congress aboard the shuttle: Utah Senator Jake Garn, and Florida Congressman Bill Nelson. Because of their NASA oversight roles in congressional committees, the agency thought it appropriate to take these lawmakers to orbit to see firsthand how the taxpayers' funds were being spent. Critics thought their flights had as much to do with greasing the skids for future budget requests as with keeping Congress informed. After *Challenger* was lost, NASA suspended this category of flight opportunities.

John Glenn's case was a hybrid—he was a proven astronaut as well as a politician—and in a way, we at NASA viewed his flight as a collective "thank you" for his bravery and skill back in the early days of the Moon race.

Dr. Jones's Corner

Glenn's flight gave the young engineers and scientists at NASA a real connection to their history, and my colleagues who flew with him had the privilege of working in space with their childhood hero. It was a magical moment in NASA's story.

Tourist in Space

2001 was not only the name of Arthur Clarke's famous space novel, it was also the year of the first *"space tourist* odyssey." And it was a strange venture.

Dr. Jones's Corner

I saw for myself in February 2001 just how demanding the workload was on the crew of the ISS—they were putting in 18-hour days, and staying up well into their sleep periods to operate and maintain the growing outpost. I thought Tito's visit would be at best an unneeded distraction, at worst a safety hazard.

The Russian Space Agency (RSA), so strapped for funds it could barely produce the boosters and space-craft it had promised for the International Space Station (ISS), was engaged in a desperate search for cash. (We'll get into the details of the new space station in Chapter 27.) The Russians were quick to notice that they could fly their Soyuz to the station with just two cosmonauts, leaving a third seat free for a paying customer. Doing some discreet marketing, they found a willing candidate in American engineer and millionaire Dennis Tito.

In mid-2000 Tito signed a contract with RSA to fly on a quick visit to the Mir space station, an opportunity that vaporized when the Mir program ran out of money, and the 15-year-old station burned up on re-entry early in 2001.

Tito, who was nothing if not a sharp businessman, pressed the Russians to take him instead to the ISS. RSA agreed to fly him there in April 2001. His trip erupted in controversy when NASA and the other ISS partners objected to a tourist visit in the middle of a critical series of assembly operations at the ISS.

RSA still had a contract to deliver on, and they really needed the cash to keep their Soyuz assembly line open. In the end, NASA received assurances that Tito would stay out of the way of the ISS crew, supervised by his Soyuz colleagues. The ISS partners reluctantly gave their permission for a visit that was going to come off with or without their approval, in return for a promise by the Russians to meet new standards for paying visitors in the future. Tito's visit to the second station crew in April 2001 highlighted some tensions with our Russian counterparts, but did force NASA and its partners to get ready in a systematic way for the opening of space to the tourism industry.

"It was paradise, a great flight and great landing, there were absolutely no difficulties" Tito said after his return. "It was perfect. I had my dream." No doubt about it, space tourism is a reality, and it's a good and necessary development for the future of commercial work in orbit.

The first tourist in space illustrated some of the growing pains NASA was experiencing with its most ambitious project since Apollo—the construction of a huge research outpost in orbit and the start of permanent human habitation in space. The International Space Station had been a long time coming, and its growth couldn't help but reflect its troubled development history.

The Least You Need to Know

- ◆ The space shuttle came into its own in the 1990s as a versatile science platform and space probe launcher.
- ◆ NASA's renewed emphasis on safety resulted in more than a hundred successful shuttle flights by the turn of the century.
- ◆ The shuttle launched three Great Observatories—the Hubble Space Telescope, the Compton Gamma-Ray Observatory, and the Chandra X-Ray Observatory.
- ◆ John Glenn finally got his second, well-deserved spaceflight in October 1998.

Chapter 27

The International Space Station—Call Sign "Alpha"

In This Chapter

- ◆ The modular space station
- ◆ Piece-by-piece construction
- ◆ Installing a new front porch
- ◆ Largest orbiting object ever

In Chapter 1, "My Adventures in Space," I told the story of my most recent shuttle mission, bringing the Destiny laboratory module to Alpha (also called ISS). In fact, NASA's human spaceflight program has been focused since late 1998 on the construction of history's largest space station. Nearly every space shuttle mission through 2006 will be headed to the new outpost, a hoped-for stepping stone to destinations beyond.

Reagan's Dream

President Reagan kicked off the project in 1984 when he called for a new space station, Freedom, to be built within the next ten years. NASA soon lined up the European, Canadian, and Japanese space agencies as partners, and planning began in earnest. However, the cost estimates soon spiraled well beyond the $8 billion price tag NASA had signed up for.

Trying to limit the cost growth, the agency shrank the size of the outpost, and postponed its construction into the mid-1990s. That date was a slippery goal. I remember how for several years, as we discussed the station in Astronaut Office meetings, the start of construction was delayed time and again. The first launch never seemed to get any closer.

After a series of cost-saving redesigns, the program was on the verge of cancellation when the new Clinton administration came into office in 1993. The station was still beset by money problems, and no one in the new administration seemed to be interested in seeing it built.

Russian Partnership

The end of the Cold War proved to be the station's salvation. Looking for a way to divert Russia's technical personnel from going to work for the highest (and potentially most hostile) bidder, the administration and NASA came up with a plan to bring Russia onboard the station program as a major partner. In exchange for $400 million in U.S. funds, the Russians would invite U.S. astronauts aboard their Mir space station. As full partners, the Russians also agreed to furnish major space station components as well as the cargo boosters for lofting supplies and rocket fuel to the outpost. While giving Russia's collapsing space program a healthy shot of cash, NASA hoped it would benefit from their experience and from their ability to loft heavy cargo into orbit cheaply.

With Russia now aboard, the multinational project became known, appropriately enough, as the International Space Station; it took on the Alpha call sign informally in the mid-'90s, and the first station commander, Bill Shepherd, used the name in space when his crew came aboard in late 2000. The station, which has been under construction since 1998, consists of a series of modules that have been carried one by one into space by the space shuttle or Russian Proton boosters. Most of the construction uses robot arm and space walking techniques honed during shuttle missions over the last five years. My canceled 1996 space walks were aimed at practicing those skills, and in February 2001 I finally had the honor of visiting and helping expand the station. About 35 missions spread out over seven years will be needed to complete the structure.

The Dream Realized

When Alpha is completed—with luck, by early 2006—it will have an end-to-end width (wingspan) of 356 feet, stretch 290 feet long, and will tower 143 feet tall. It will have a mass of nearly one million pounds and will provide living space for up to seven astronauts and scientists. The pressurized living and working space aboard the completed station will be more than 46,000 cubic feet, roughly equivalent to the passenger cabin volume of two

747 jetliners. Indeed, when the station is complete, it will be, by far, the largest and heaviest object ever to orbit the Earth. The station will be the size of two football fields laid side and by side, and will be clearly visible from Earth, appearing as a bright star crossing the night sky. In fact, it's easily visible right now, even in the glare from city lights. (Check the time of the next ISS pass over your house at www.spaceflight. nasa.gov.)

Cosmic Facts

Each space station crew has at least one person who has received medical training—like an emergency medical technician—to handle minor injuries and illnesses using the station's medical kit.

Sunrise, Unity, and Star

The first piece of the ISS, built by Russia but paid for by the U.S., is called the Zarya (Sunrise) module, and it was lofted into orbit in November 1998 by a Russian Proton rocket. Just a few weeks later, in December 1998, Space Shuttle *Endeavour* delivered the first American-built chunk of the ISS, Unity, to orbit. Unity has six docking ports and was attached to Zarya using *Endeavour*'s robot arm. The third element, the Russian service module Zvezda (Star), was launched in July 2000. Zvezda, also called the Service Module, is an updated version of the core module from the old Mir station. It furnishes propulsion, life support, and living quarters for the crew for the first several years. The Z-1 truss (housing attitude control gyros) and first giant set of solar panels were added to the space station in fall 2000. The first crew arrived to take up residence in November 2000.

Cosmic Facts

The ISS draws upon the resources and the scientific and technological expertise of 16 cooperating nations, including the United States, Canada, Japan, Russia, and 11 participating member nations of the European Space Agency (Belgium, Denmark, France, Germany, Italy, The Netherlands, Norway, Spain, Sweden, Switzerland, and the United Kingdom). In addition, Brazil and Italy have signed on as hardware and payload contributors.

That first crew was composed of Commander Bill Shepherd, a U.S. astronaut; Soyuz Commander Yuri Gidzenko, a Russian cosmonaut; and Flight Engineer Sergei Krikalev, a fellow cosmonaut. Shepherd's crew worked with us closely in activating and outfitting the Destiny lab when we brought it up to the ISS. They were launched on a Russian Soyuz spacecraft on October 31, 2000, and returned on the shuttle four and a half months later. NASA hopes that we'll never see a future time when human beings will not be in space.

Politics

One of the biggest challenges facing NASA in leading the station program has been dealing with the different interests of the nations involved—and not just the differences between Russia and the United States, but those between all the partner countries. For example, to boost support at home, each major partner has pushed to have its astronauts assigned to shuttle assembly missions and long-duration station crews as soon as possible; juggling those competing demands has made crew selection in Houston a political football at times.

Money

But by far the largest headache for NASA has been the precarious state of Russian finances. Instead of saving the U.S. money by contributing station hardware and booster capacity, the Russian Space Agency (RSA) has instead needed transfusions of American funding. Starved of money from its government, RSA delayed the launch of its Service Module by two years, costing NASA millions of dollars as our finished station components sat in storage at KSC. To be fair, American hardware was seldom ready to go on time—the Destiny lab was nearly as late as the Service Module—but the constant Russian delays threatened to erode confidence in their long-term commitment to the ISS.

Cosmic Facts

The concept of a space station was first put forth by writer Edward Everett Hale in 1869. He called the station a "brick moon."

As of this writing, it appears the Russians are meeting their cargo and Soyuz rescue vehicle commitments, and the level of cooperation in operating the station day to day is growing steadily. The money problems are still there on both sides, but it's evident that both have a strong willingness to work together technically, and in seeing Alpha succeed.

Partnership

More than a few of the ISS's technical problems stem from coordinating the station's far-flung partnership. Dozens of NASA engineers and astronauts moved to Moscow to coordinate the details of how the Russian components would mesh with the rest of the station. Everything from how to make electrical connectors compatible, to what English labels should go on the control panels in the Russian-built modules, had to be worked out in a never-ending series of detailed, and sometimes acrimonious, technical meetings.

This technical attention to detail is a tough business, and we still don't have perfect coordination with our Russian partners. On my flight to the station, for example, we carried a large spare part to orbit—the Vozdukh carbon dioxide scrubber. This Russian-built assembly was supposed to replace the identical one originally launched in the Service

Module. Russian engineers had packed the Vozdukh in a cube-shaped aluminum box about the size of a large television. They'd checked its snug fit into its launch slot in *Atlantis*'s middeck and made sure (they assured us) it would pass through the various orbiter and station hatches on its way to the Service Module.

Once docked to ISS, we got the aluminum box through the shuttle airlock's two hatches, around the spacesuits stowed there, and headed up to the docking tunnel hatch into the station. Yuri Gidzenko and Sergei Krikalev aligned the box carefully with the opening, glided upward and—thunk!—the box was an inch too big. We had to float the container down into the middeck again, and strap it out of the way until the ninth day of the flight, when Yuri took the box apart with his tool kit, and brought the components through one by one. The Russians had done the hatch fit check on a mockup—close, but not identical to, the real space station hatchway.

Cooperation

Fortunately, culture and politics have much less of an effect on the international crews that actually occupy the space station. Those space teams work as a single unit despite language and cultural differences. So far, all the long-duration crews have been Russian and American teams, and the professionalism and common experience of astronaut/cosmonaut training has forged strong bonds among them. The crewmembers all share common goals: establishing a permanent human presence in space, bringing research benefits from space down to Earth, and gaining the experience on long-duration flights to enable human missions to the planets.

NASA hopes that the cooperative success of the space station crews, despite their different cultural and spaceflight backgrounds, will lead to cooperative efforts to explore beyond Earth's orbit. NASA also hopes that that sense of cooperation in space will in some small way enhance international relations on Earth. Certainly we will need to cooperate with other countries in facing other technical and scientific problems around the globe—conserving resources, global climate change, pollution controls—and the ISS effort will serve as an example of what can be accomplished in a climate of determined cooperation.

NASA-Mir

During the late 1970s and early 1980s, the Russian space program was geared around long-duration flights aboard a series of space stations. Salyut 6, and later Salyut 7, hosted a series of cosmonaut teams who achieved impressive long-term missions in orbit, one lasting almost eight months.

Space Talk

Mir means both "peace" and "world" in Russian.

In February 1986, the U.S.S.R. launched the space station *Mir*. Boosted by a Proton rocket, this station had more electrical power and more automatic life-support systems than the two previous stations. Mir expanded to include several scientific modules, larger living quarters, and a forest of solar arrays. The spidery space station was a showcase for the expertise the Russians had developed in operating and maintaining a complex machine in orbit for years.

Following the fall of the communist Soviet Union, there was a renewed cooperation between the U.S. and Russian space programs. This cooperation culminated in the invitation to the Russians to join the ISS program (and at the same time save it from cancellation). As a result of that agreement, by 1995 the U.S. space shuttle was docking with Mir on a regular basis. Russian cosmonauts were going into orbit aboard the shuttle, and American astronauts were being launched aboard the Soyuz.

Seven Americans lived aboard Mir, each paired with two Russian cosmonauts. Norm Thagard was the first American to ride the Soyuz to orbit; his pioneering flight was an exercise in sheer stamina. He lost about 20 pounds, and because of the language barrier suffered from isolation. Shannon Lucid set a new American record by remaining in orbit for nearly six months. John Blaha's flight continued the scientific research aboard Mir. Jerry Linenger lived through a dangerous fire that threatened to burn through the Mir's pressure hull. Mike Foale survived a collision between Mir and its Progress cargo ship that depressurized a module to vacuum. Dave Wolfe helped life get back to normal aboard Mir, and Andy Thomas successfully ended the program with a long stay aboard in 1998.

Getting an early start on working together was essential—the ISS will be four times larger than Mir when its construction is completed. Without the experience gained aboard Mir, and the relationships forged during those seven flights, the planning and construction of the ISS would have been far more difficult, if not impossible; it takes time to build confidence and relationships with the sometimes-less-than-forthcoming Russian space program.

Progress Cargo Ship

Just before the arrival of the first crew in late 2000, Alpha—like Mir before it—began receiving its supplies through periodic visits from the Progress, a Russian unmanned space vehicle that functions as a cargo freighter for the space station. Similar to the Soyuz, but without a re-entry capability, the cargo ship carries as much as two and a half tons of supplies on each trip.

In fact, when Mir's mission was complete and it was time to de-orbit the station, it was a Progress freighter that docked with the space station and performed the engine firings that brought it down into the atmosphere, where both burned up.

> **Cosmic Facts**
>
> The first mention of a space station in popular literature came from Kurd Lasswitz, a founder of science fiction, in 1897, who said that space stations would be necessary for future space travel.

Destiny in Space

In February of 2001, I flew on my last mission, aboard *Atlantis*. We installed the U.S. Laboratory module onto the front end of Alpha, which officially made the space station the largest spacecraft in history.

The Destiny laboratory module is the nerve center of the ISS, with provisions for computer command and control, life support, power distribution, and of course, scientific research. After installing and activating the lab, we turned it over to the Expedition One crew, who were delighted to see their work and living space grow by about 30 percent. Station operations have now shifted from the Service Module to the Destiny lab, with Houston taking over the role of lead control center. (See Chapter 1 for more about this mission.)

The Robotic Arm

After *Discovery* returned the Expedition One crew in March of 2001, the STS-100 crew on *Endeavour* docked with Alpha, this time to install the station's very own robotic arm. This 58-foot-long arm would be needed shortly to continue the station's construction (out of the reach of the shuttle arm) and to perform planned and emergency repairs.

Dr. Jones's Corner

Every time the space shuttle has a mission, it returns with many small pockmarks in its protective thermal tiles (and sometimes its windows and cargo bay door radiator panels), a result of bombardment by space debris and micro-meteorites. On a couple of my flights, I've seen pits in our windows about the size of a pinhead, surrounded by a little halo of debris blown out of the tiny crater. Over the last several years, hundreds of impacts have been documented and repaired. For its 15-to-20-year mission, vital areas of the ISS are protected by aluminum and Kevlar debris shields that blanket the actual aluminum pressure hull (itself about ¼ inch thick). On my STS-98 space walks outside the station, I observed impact damage on one of the small radiators used to keep electronic equipment cool. Over the years, astronauts will certainly have to repair damage caused by these hypervelocity collisions.

Because of the central role the Space Station Remote Manipulator System (SSRMS) would play, this mission was crucial to the future of the program. If the robotic arm didn't work, the rest of Alpha's construction would be held up until it was repaired or replaced.

Cosmic Facts

The robotic arm delivered on mission STS-100 was built in Canada and is known as Canadarm 2. It's the major Canadian contribution to the ISS program. The Canadians had done their job well. Canadarm 2 was soon perched on its base on the hull of Destiny, the same fixture that Bob Curbeam and I had installed outside in February. After some initial checkout problems with the backup commanding circuits, the arm was ready to take the handoff of the next major component coming up—the airlock module.

New Airlock

Alpha's July 2001 home addition involved the installation of an airlock, which has been called the "world's most expensive front door." It's true that the six-and-a-half ton airlock cost $165 million, but it's not just a door. It's also a front porch.

Looks Like a Perfume Bottle

The 18-foot-long and 15-foot-wide airlock—which resembles a perfume bottle with a broad base and a slender neck—works like a pressure valve. Air pressure in space is zero, and inside the ISS it is 14.7 pounds per square inch. To perform a space walk, astronauts go from the station into the airlock, closing the airtight hatch behind them. The airlock will then have all the air pumped out of it until it is completely depressurized. The astronaut can then safely open the outer door and float out into space. When they want to return to the station, they go back into the airlock, which is then re-pressurized. Once this is done, the astronaut can then re-enter the station.

Before the new airlock was installed, EVAs could only be performed when the shuttle was docked to the station (the shuttle has its own airlock), or by using Zvezda's emergency airlock. Now, the station crew can go on an EVA as needed to repair the station or assist in assembly. The new station airlock is compatible with both U.S. and Russian space suits. Like the shuttle airlock, astronauts can drop in during an EVA to recharge their suit's batteries or supply of oxygen.

Dr. Jones's Corner

Before any station assembly mission, like the one to deliver the airlock, *Atlantis* first had to dock with the space station. Rendezvous usually occurs about 48 hours after launch, on the crew's third workday in orbit. Shuttle commander Lt. Col. Steven W. Lindsey guided *Atlantis* in for the 230-mile-high docking a few minutes late, taking extra time to line up the spacecraft as perfectly as possible. Lindsey had to keep the orbiter's docking system lined up within plus-or-minus three inches of the station's centerline target and made the final deliberate approach at 0.1 feet per second. When the two docking rings make contact, the shuttle's thrusters automatically fire to nudge the two craft firmly together, and latches on the orbiter docking system snap over the station's docking ring for initial capture. Inside the cabin, we hear a loud bang, a series of cannon-like thumps from the thrusters, and shouts of "Capture!" from the flight deck crew.

Shuttle commander Steven W. Lindsey's five-person crew had the job of delivering the airlock, on its first trip to space since taking Destiny to the ISS in February. They blasted into space from Cape Canaveral on July 12, 2001, at 5:04 in the morning, hitting a launch window that was only about five minutes long. Once docked, the Expedition Two crew lent a hand with airlock installation. The plan called for using both the shuttle's robotic arm—the same type I used to wrestle the *Wakeshield Facility* out of space—and the station's advanced Canadarm 2, added to the station in April.

In a series of three EVAs, mission specialists Michael L. Gernhardt and James F. Reilly helped line up the airlock and install its high-pressure nitrogen and oxygen tanks. On the third space walk, they used the new airlock for the first time to go in and out of the space station. Air Force Colonel Susan B. Helms, my classmate and a member of the Expedition Two crew, operated the space station's robotic arm—she took the handoff of the airlock from the shuttle's robot arm and berthed the new module aboard the ISS.

The Alligator and the Snake

Before the airlock left *Atlantis*'s cargo bay, Gernhardt had to remove and stow a large thermal cover protecting the airlock's berthing ring. Manhandling the stiff cover into a bundle, he then had to wrap it up securely with several tethers. "It was like wrestling a 12-foot alligator and tying it up with a 20-foot snake," Gernhardt said. Once the airlock was securely in place, the two crews held an impromptu dedication ceremony, stretching a piece of white paper across the entrance hatch for the lighthearted ribbon cutting.

Space Talk

The ISS's new front door is called the **Joint Airlock Module.** The airlock was manufactured by Boeing Company at the Marshall Space Flight Center in the same building where the test version of the Saturn V rocket's first stage was built.

Human Research Facility

Continuing the outfitting of the station, the following shuttle mission, STS-105, headed to the ISS in August 2001. In the Italian-built cargo module in the payload bay was a new experiment rack—the Human Research Facility (HRF). This biomedical laboratory is designed to study the human body in space. The HRF will have its own slot inside Destiny—it has its own computers and diagnostic equipment so that it can measure different changes within the body during these multi-month missions.

Space Talk

Cardiopulmonary means having to do with the body's heart and lung system.

Cosmic Facts

Since the arrival of Destiny, the third station crewmember has been sleeping in an empty experiment rack in the lab. When the Italian-built Node 2 arrives in 2003, it will house four individual crew quarters, each with its own sleeping bag, storage bins, reading light, and ventilation fan. The crew quarters will eventually be moved to the U.S. habitation module, the last major element scheduled to arrive at ISS in 2006.

There is one experiment that measures the *cardiopulmonary* responses of a space traveler, and how those responses change over a time in free-fall and during EVAs. For example, the lab will measure changes in the astronaut's lung capacity—the amount of oxygen taken into the body with each breath. That information can be stored in the HRF and later relayed to the ground.

Tough Duty

The first astronauts and cosmonauts to occupy Alpha had to endure more than a few hardships. Like in any new apartment, there were some glitches that needed to be worked out, the job getting easier with growing experience and skill.

The first three astronauts to live on the station were forced to wear ear plugs when sleeping because of noisy machinery in the Service Module. In fact, Zvezda has only two sleep stations—Yuri Gidzenko had to sleep in the "hallway" of the Zarya module, tethering his possessions and sleeping bag to the wall. The offending machinery: an "air scrubber" designed to remove carbon dioxide from the air. The machine had the annoying habit of emitting a loud blast of sound every 10 minutes, just frequent enough to make sleeping extremely difficult.

Struggle Is Worth It

Privacy was a rare commodity—bathing and bathroom chores all take place in the tight confines of the Service Module. Trash containers, food supplies, and spare parts partially filled the passageways. Small tools and parts would go missing for days. The U.S. carbon dioxide scrubber broke down upon activation. The astronauts were given a DVD player, but the screen was too small for all to see comfortably. Annoying computer glitches were all too frequent. Even the laptop that was supposed to record caution and warning alarms broke down.

Exploration and the search for new knowledge has never been easy, and life on the frontier is tough on the pioneers. But the station keeps getting bigger and better, and all of the troubleshooting is definitely worth it. Why? I'll let another astronaut explain.

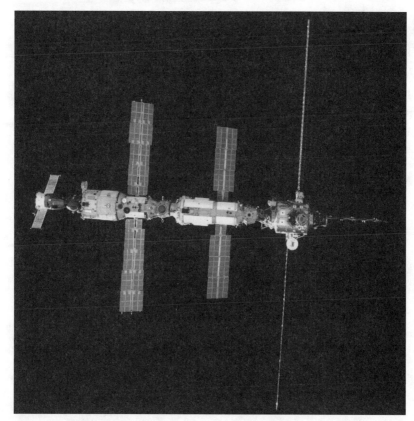

The International Space Station as it appeared in February 2001.

(NASA)

According to Expedition Three Commander Frank Culbertson:

> We need a space station because we need a frontier. We need to keep pushing the human race to expand beyond the current boundaries that we have. Throughout history those boundaries have been a combination of physical boundaries, such as oceans, and mental boundaries, that feeling of, "Can I really get to the other side of that ocean?" We still have mental leaps we must make when it comes to exploring space. And I am absolutely 100 percent confident that we will wind up on the other side of that boundary. We will make space a more comfortable place in which to live, work and even visit.

The Least You Need to Know

- The ISS is known by its crews as Alpha.
- Alpha is the largest object ever to orbit the Earth.
- The station is being constructed piece by piece by a series of space shuttle and Russian Proton and Soyuz flights.
- Although a spartan facility, Alpha holds the key to our future in space research and our long-term hopes for human exploration of the solar system.

Chapter 28

How a Space Mission Gets Off the Ground

In This Chapter

- ◆ The birth of a mission
- ◆ Training the crew
- ◆ Dress rehearsals with Mission Control
- ◆ Safety from the bottom up

How does a space mission get off the ground? The process begins at least a year before flight, and sometimes as much as five years before. A mission is born through a combination of two plans—those of the space shuttle program, and those of the payload "customer."

Behind the Scenes, Department by Department

The space shuttle program, headquartered at the Johnson Space Center in Houston, Texas, develops a flight schedule that extends at least five years into the future. In addition to designating launch dates, orbiters to be flown, and the order in which they're to be launched, the schedule establishes a practical turn-around interval between each flight. This shuttle flight schedule combines input from the Mission Operations Directorate at JSC (How many

Dr. Jones's Corner

As you might guess, the budget up at NASA Headquarters in Washington determines how many flights we can launch per year.

crews and flight controllers can be trained each year?) and from the Kennedy Space Center (How quickly after landing can the work force at the Cape return a space shuttle to launch readiness?).

In the space station era, the major payload "customer" is the space station program, also centered at JSC. The International Space Station (ISS) program has developed, with its partners, an "assembly sequence" that shows how and in what order the station gets put together. My 2001 flight, for example, was ISS assembly flight 5A, the fifth American mission hauling a major component to the station. In early 1997, a couple of years before the hardware was needed in orbit, the ISS program asked the shuttle program to put this mission on the flight schedule.

Today, in the same way, the space shuttle program tells the ISS program how many flights per year are planned, and how much mass those shuttles can carry to orbit. The space station managers then identify which ISS modules or cargo they wish to fly on those shuttle flights. Based on those requests, the space shuttle program creates a flight manifest which matches an orbiter and a mission number (STS-"XXX") with a payload—the U.S. Destiny Lab, for example (see Chapter 27, "The International Space Station—Call Sign 'Alpha'").

Once the shuttle program has firmed up its flight manifest, it assigns a Flight Manager, and asks the Mission Integration Office at JSC to pull the basic elements of the flight together. A Flight Integration Manager (we call them "FIMs") coordinates an analysis of whether a mission is feasible, given payload, orbiter lift capacity, and the laws of physics. Now flight design can begin.

Flight Design

The *MOD* takes on the task of flight design once the payload and orbiter have been matched.

This is where old Isaac Newton comes in again. The flight design team takes the desired orbit, the weight of the payload and orbiter, mission duration and launch date, and, using basic orbital mechanics, creates the physical parameters of the mission. From their analysis comes the time of day of the launch, the weight of propellants needed on the orbiter, the amount of breathing oxygen and nitrogen that must be carried, the length of time the orbiter can stay in orbit with its load of fuel cell reactants, and the proper time to deorbit for a landing at the Cape.

Space Talk

MOD stands for mission operations directorate.

Most important from my point of view—the flight design determines how many crew-members are needed, and how much weight to orbit (the "ascent performance margin") is available to actually fly them! Sometimes the flight is very similar to a previous flight—so much of the information can be pulled from that experience. The flight design summary is the blueprint for the rest of MOD in working up a flight plan.

Mission Integration

With the basic flight design in hand, the FIM brings together all the organizations involved in preparing for and planning the flight. He or she sets a number of planning milestones, with the details of the flight steadily firming up until the plan and the hard-ware come together at the Cape at T-minus-zero. The FIM works directly for a space shuttle program Flight Manager, responsible for making sure the mission is ready to go on time, with all NASA organizations ready to support the flight.

Product Teams

The Flight Manager and FIM regularly convene all the mission participants, including those at the contractor and the Cape, to develop the plan and identify and fix any problems that crop up. This group is known as the *IPT*.

Space Talk

IPT stands for Integrated Product Team. The IPT serves as a troubleshooting team that has the horsepower to make decisions and review the performance of all the key organizations. The managers at the IPT identify any problems that they're running into, and the other IPT members can commit on the spot to fixes or solutions as they're proposed. The next week, the team reconvenes to gauge progress. There are also smaller IPTs that support the Mission IPT: I was heavily involved with our EVA IPT, in charge of making sure all our spacewalking tools and suit gear were ready in time for flight.

I became a member of the STS-98 IPT when I was assigned to the flight in June 1997. The IPT meetings taught me the basics of the flight design—which vehicle I'd fly in, how long I'd be gone, when our launch date was, and some basic facts about our payload, the Destiny Lab. Most important, I met the people who could tackle problems or evaluate ideas or suggestions from the crew.

Crew Assignment

One of the burning questions for any astronaut, which pops into one's mind about, oh, a dozen times a day, is "How do I get assigned to a mission?" It's a mysterious process for the astronaut, and for years I couldn't make much sense of it, but here's my sense of how it happens.

When the flight design takes shape, MOD (Mission Operations, which is also responsible for training) estimates how long it will take to train a crew for the mission.

MOD takes its training estimate and formally notifies the Flight Crew Operations Directorate (yes, that's FCOD, which includes the Astronaut Office) that NASA needs a crew assigned. The selection process is then in the hands of the Chief, Astronaut Office—my old boss.

Cosmic Facts

The Chief also strives to reward good performance in ground assignments with the privilege of flying in space.

Cosmic Facts

Soon after "the phone call," the assignment news is made public; the new crew's first priority is hosting a beer bash at one of the favorite watering holes along NASA Road One in Houston.

My Air Force Academy classmate Charlie Precourt, a retired U.S. Air Force Colonel, is currently chief of the Astronaut Office. Armed with the flight manifest and the MOD training schedule, Charlie sharpens his pencil, sits down, and takes a look at his personnel roster. The Chief has a lot of balls to juggle. He needs two pilots up front, a mission specialist as flight engineer, and usually at least two more mission specialists to handle the payload duties. He needs a mix of veteran astronauts and rookies eager to gain experience. And he needs a crew with the right mix of skills to pull off the job they're going to be given—robot arm experience, perhaps a veteran space walker, a commander who's already been part of a previous shuttle rendezvous.

Charlie pencils in the crew and sends the list up to his boss, the Chief of Flight Crew Operations; in late 2001, the Chief of FCOD was veteran mission specialist Steve Hawley, hired by NASA with the first batch of shuttle mission specialists in 1978.

With the Chief of FCOD's approval (and any higher OK's he needs), the Chief Astronaut makes the crew assignment official by calling each astronaut in to give them the good news. Sometimes the call comes from the FCOD Chief, but most of the time, Charlie gets the honor of bringing a long-time dream to reality for a rookie or giving a veteran astronaut a new and exciting challenge. The crew commander consults with the Chief about how to assign the mission tasks among the various crewmembers, but in my experience, the final details are best worked out among the crew.

Mission Operations

The person responsible for accomplishing the mission objectives is the Flight Director. He or she puts together the crew, a team of flight controllers, instructors, and their supporting engineers, and focuses these dozens of people on what it's going to take to achieve success. In running the mission, the Flight Director has all available information at his or her fingertips. While the crew commander has responsibility for the immediate safety of the spacecraft and crew, and the execution in orbit of activities from the flight plan, the Flight Director makes the operational decisions that guide the larger course and strategy of the mission.

> **Cosmic Facts**
>
> For STS-98, the mission to install the Destiny laboratory, we had a specialized flight director, Leroy Cain, who handled the dynamic phases of launch and landing, and a lead Flight Director in charge of the overall mission. In the latter role, Bob Castle's command of detail and encyclopedic knowledge were beyond impressive.

Flight Plan

MOD begins to develop the detailed flight plan for the mission around a year before launch. The Flight Activities Officer, or FAO, first identifies the major activities of the flight. Using a set of scheduling rules that prevents overworking the crew, and safeguards the minimum sleep requirements, he develops a summary of the flight plan. For example, on STS-98, the flight plan gave the crew two nights of sleep in orbit before attempting the crucial rendezvous with ISS, and put the first assembly space walk four days into the mission, long enough for Bob Curbeam and I to have gotten our space legs. FAO gave us a half-day off duty just after undocking from ISS and made sure that the pilots got plenty of sleep on the night before landing. The flight plan bounces back and forth between FAO, the crew, and the flight control team until all are satisfied that we have a workable game plan.

The final version of the flight plan breaks the mission timeline down into five-minute increments, and each crewmember can tell what activity they should be involved in down to the minute. The flight plan refers to events in "mission elapsed time," the time in days, hours, minutes, and seconds since lift-off.

Training

MOD quickly meets with the new crew and assigns them a training manager and an instructor team. I can't overemphasize how vital this step is to the success of the mission. The instructors, all experts in space shuttle and space station systems, take on the job of teaching the crew, and then rehearsing them until they can cope with the mission objectives and any emergencies that might throw them off track.

Over the course of a year or more, the training team and the crew forge a tight bond—the crew learns the shuttle and station systems cold, and the training team learns the strengths and weaknesses of the astronauts. The instructors focus their efforts at shoring up those weak areas and making sure that the teamwork is in place to deal with the worst problems the spacecraft or mission can throw at them.

Simulations

After initial classroom training in the spacecraft systems, the crew begins simulator training about seven months prior to the launch. First, it's just the instructor team leading the crew through launch, entry, rendezvous, or docking. The Simulator Supervisor ("Sim Supe") and the instructor ("team lead") develop simulator scripts with systems experts, and each time the crew steps into the simulator, the team plays out an emergency scenario designed to show them the most likely—or most dangerous—malfunctions they could encounter. This "stand alone" training brings the crew up to basic proficiency in their mission skills, ready for more complex and subtle failure scenarios.

This final phase of spaceflight training blends the crew in the simulator with the team of flight controllers in Mission Control. For the crew, the high-fidelity shuttle simulator behaves as if it were really in orbit: The Earth wheels by outside the windows (high-definition TV screens), all switch actions and computer keystrokes produce a realistic effect on the vehicle, and control inputs produce the thump of a thruster firing, as well as the realistic visual impression that the shuttle is pivoting or translating through space.

> **Cosmic Facts**
>
> The rehearsals reach a climax less than a week before launch, when the crew grabs one last simulator session to run through launch and landing failures. We were already in medical quarantine by then, tired of the training, but convinced we were ready to handle anything. Our experience on STS-98 proved the unexpected is just what is likely to show up.

Into this mix, the instructor team tosses a volatile mix of system failures, nuisance alarms, and insidious malfunctions, doing their best to overwhelm the flight crew and the flight controllers. The effect is so realistic that my mental frame of mind shifted past the fact that I was in a simulator in the box-like Building 5—I was actually on the orbiter flight deck or in the airlock. And the flight controllers linked to us by radio were not a few hundred yards away across campus, but were observing our actions and helping us out from the other side of the globe.

In a concentrated burst of simulations in the last few weeks before launch, the crew, flight director, and flight controllers come to work as a well-practiced team, focused on the mission objectives. On my last mission, we saw so many failure scenarios, and our jobs were so harried by malfunctions, that we became used to the constant pressure and obstacles.

Flight Readiness Review

The last hurdle the team has to clear before launch is the Flight Readiness Review. Held at Kennedy Space Center by senior shuttle, station, and Cape managers about a week before launch, the review airs out all the remaining issues on the table. Safety gets first priority—any solutions for a given technical problem get approved only when the team agrees that safety of flight will not be compromised. I've had my flights postponed at the Flight Readiness Review for main engine inspections, for worries about booster nozzle flaws

> ### Dr. Jones's Corner
>
> The usual training time for a mission is about a year, but for particularly challenging and complex flights, some crewmembers need to begin preparing as much as two years beforehand. (I was assigned to STS-98 about two years prior to the planned launch, to assist with the planning of our three complex EVAs.)

seen on a previous mission, and for unresolved questions about the safety of wiring on the outside of our solid rocket booster casings. The legacy of *Challenger* is still alive and strong—we don't launch until all agree it's safe to fly.

So now you can see why I recommended playing team sports to prepare yourself for a career with NASA. In order to get a space shuttle mission off the ground, many people must get together and share their knowledge and experience. It is also important to understand that no NASA mission occurs in a day, or a week, or a month. Each mission, whether it be the space shuttle or a robotic satellite, goes into the planning stages years before the launch.

The Least You Need to Know

- A space shuttle mission is created by matching a flight opportunity with a major payload customer.
- Flight design typically begins about a year before launch.
- The Chief Astronaut picks a crew about a year before flight.
- Simulator training prepares a crew and the flight control team to handle any credible failure scenario—and the unexpected.
- Safety has the final word at NASA's Flight Readiness Review.

Chapter 29

Revealing the Planets: Today's Robotic Explorers

In This Chapter

◆ Looking outward

◆ Exploring the solar system—and beyond

◆ Planets, moons, and asteroids

◆ Catching a piece of the Sun

The Apollo moon landings would have been even riskier than they were without the help of robotic explorers like Ranger, Surveyor, and the Lunar Orbiter. As we'll read in Chapter 30, "Our Future in Space," NASA and its partners have ambitions to leave Earth orbit once again. But we can't do it without getting some crucial missing information about our solar system and the planets through robotic exploration.

We've learned a tremendous amount about our neighboring planets in the past 40 years, all because of our robotic probes. It's been the only way to proceed, because of the great distances and hazards involved in interplanetary travel. And, for the time being, that's probably going to remain the case. So our robot investigators are more important than ever. In fact, they're becoming ever more ambitious and capable. A robot craft landing on Mars or Venus can tell us the exact temperature of the planet, the composition of its atmosphere, and take crystal-clear full-color photographs of the surroundings. A radar mapper can reveal the shrouded surface of Saturn's moon, Titan. A robot will help us get our hands on pieces of a comet.

And so, in this chapter, we look at NASA's ongoing program of robotic exploration, and the machines that replace our eyes and other senses as we dig deeper into the mysteries of our universe. Perhaps the premier robotic extension of our senses has been a sophisticated piece of optics called the Hubble Space Telescope.

Hubble Space Telescope

Scientists have long known that they would get a better view of deep space if they did not have to view it through the dust and distortion of Earth's atmosphere. A telescope in space would give us a superb vantage point for taking a detailed look at our galaxy and beyond.

Astronomer Lyman Spitzer proposed orbiting such a telescope in 1946. His dream was realized by the Hubble Space Telescope, an orbiting spacecraft which contains a nearly eight-foot-diameter telescope and its supporting machinery. The telescope, delayed for four years by the *Challenger* disaster, was placed into orbit by the Space Shuttle (STS-31, launched April 25, 1990).

The objectives of the HST are to …

◆ Investigate the composition, physical characteristics, and dynamics of celestial bodies.

◆ Examine the formation, structure, and evolution of stars and galaxies.

◆ Study the history and evolution of the universe.

◆ Provide to astronomers a long-term space-based research facility.

Cosmic Facts

Edwin P. Hubble, for whom the space telescope is named, was the astronomer who used the large telescopes at Mount Wilson Observatory in California. He discovered that the Milky Way, which had appeared to be sufficiently vast to contain everything in the universe, was only one of many galaxies. Hubble was also the first to theorize that the universe of many galaxies is not static, but expanding. The universe—time and space—had a definite beginning (the Big Bang) and has been growing and changing ever since.

When first deployed, the HST could not focus properly due to a built-in flaw in its main mirror. This embarrassing (and expensive) problem was fixed in spectacular fashion by shuttle astronauts (STS-61 in December 1993). Since that repair, the telescope has twice been serviced by visiting astronauts. Another servicing mission is slated for early 2002 and will extend the life of HST well past the middle of this decade.

Responsibility for conducting and coordinating the science operations of the Hubble Space Telescope rests with the Space Telescope Science Institute (STScI) at Johns Hopkins University in Baltimore. The Institute operates the telescope for NASA as a science facility available to astronomers from all countries.

Cosmic Facts

Scientists have used Earth-gazing satellites to photograph more than just weather patterns. Surveillance satellites are used for national security as well. Few people realize how important a role the CIA has played in our space program. Satellite reconnaissance was considered a goal of the U.S. space program as early as 1955. The Vanguard program was geared toward space surveillance, and by 1961, the U.S. was learning details about Soviet missiles through photographs taken from space.

Defense reconnaissance satellites have increased in sophistication over the last forty years. Advances in spacecraft pointing control and optics forged by the military space program were applied to the design and construction of Hubble. Hubble is a cousin, if not a twin brother, of the reconnaissance satellites of the late '70s and early '80s.

With the help of the Hubble Space Telescope, we have looked back in time 11 billion years, almost to the dawn of the universe and seen the birth of the earliest galaxies. Staring at a piece of sky "no bigger than a grain of sand held at arm's length" (as space historian William Burrows put it), HST uncovered 40 billion new galaxies, layered back in time as far as the telescope could see. Closer to home, HST has tracked storms on Jupiter, watched dust storms on Mars, and given us our best views of the remote and icy surface of Pluto.

Clementine: Back to the Moon

As part of the U.S. effort to develop ballistic missile defenses, an experimental spacecraft known as *Clementine* was launched from Vandenberg Air Force Base aboard a Titan IIG rocket in 1994. The probe, meant to test sensor and thruster technologies for missile interceptors, headed not for an enemy warhead, but for more convenient and useful targets: our Moon and an asteroid known as 1620 Geographos. Its mission was to map the Moon in even greater detail than the Lunar Orbiters and grab close-up images of Geographos's ancient surface.

Cosmic Facts

At NASA, everything gets a name. The largest rock from the lunar surface returned to Earth by Apollo astronauts weighs almost 26½ pounds. Its nickname is "Big Muley."

The craft was first put into Earth orbit, then successfully boosted into lunar orbit. After its mapping mission exceeded all expectations, plans were made to send it on to a flyby of the asteroid. However, a computer error put the craft into an uncontrollable spin, and the asteroid mission was scrubbed. Before leaving the Moon, *Clementine* discovered intriguing hints that there might be water at the lunar south pole—an important resource for future human outposts there.

Cassini Mission to Saturn and Titan

Another explorer of the solar system, and the last of the large Voyager-style probes, is *Cassini*—headed for an in-depth study of the Saturn system. Like *Galileo* at Jupiter, *Cassini* is to orbit Saturn, its rings, and moons for several years. *Cassini*, launched in 1997, is en route to Saturn now, and is scheduled to go into orbit in 2004. The spacecraft will study Saturn's atmosphere, its magnetic field, its rings, and several of its moons.

Cassini's duties include dropping the Huygens probe into the murky atmosphere of the largest moon, Titan; the probe should make it down to the surface, sending imagery back while it descends under its parachute. *Cassini* will then encounter Titan repeatedly, and map its surface with radar, much like *Magellan*'s work at Venus. Titan's frigid atmosphere may contain organic compounds and processes that can tell us about the chemical chain leading to the origin of life on Earth.

Martian Mysteries: The *Mars Observer*

There is no better fuel for the alien-abduction and UFO fire than the frankly mysterious streak of bad luck our robot spacecraft have had at Mars. Probes from both the U.S. and Russia have repeatedly disappeared while attempting to photograph or land on the Martian surface. Russia's *Mars* and *Phobos* spacecraft have met with repeated failure, often losing radio contact just as the mission was about to encounter the Red Planet. Our own interesting chapter in this saga is the story of the *Mars Observer*.

Dr. Jones's Corner

During the Cold War, the U.S. competed with the Soviet Union to win the space race. Today, with the Cold War over, the space efforts of the former Soviet Union and the United States have found many joint interests. For example, the Russian heavy-lift booster, the Proton, is marketed by Lockheed Martin, typically thought of as a U.S. defense and space contractor. And the latest version of the U.S. Atlas launch vehicle has a first stage powered by Russian Energomash engines.

Mars Observer was a NASA mission designed to study from Martian orbit the surface, atmosphere, interior, and magnetic field of the red planet. The mission was supposed to operate in orbit for a full Martian year, or 687 Earth days. *Mars Observer* would thus get a look at Mars during all four of its seasons. The spacecraft carried a high-resolution camera and remote sensing instruments aimed at determining exactly what elements and minerals the Martian surface was made of, and how they were distributed. Mars' magnetic and gravitational fields were also targets. *Mars Observer* was also to monitor the circulation of the Martian atmosphere.

Mars Observer was launched from Earth on September 25, 1992, atop a Titan 3 rocket. Communications were lost with the spacecraft on August 22, 1993, just as it was preparing to go into orbit around Mars. No significant scientific data was returned. The investigation team found that the failure was probably due to a design oversight: A fuel line explosion likely occurred when the craft pressurized its propellant tanks in preparation for the orbital insertion burn.

Cosmic Facts

Speculation on the repeated misfortunes of Soviet and American spacecraft attempting to investigate Mars became so great that *Time* magazine reporter Donald Neff jokingly proposed a culprit. He wrote in the early 1970s that the spacecraft were no doubt being eaten by a monster he termed the *Great Galactic Ghoul.*

Space Talk

No degree is required to call oneself a **ufologist,** an "expert" in the lore of unidentified flying objects.

To further intrigue *ufologists*, the last few photographs taken by the Mars Observer and beamed back to Earth appear to show an increasingly large black spot—images that could be interpreted as an object approaching the spacecraft.

Galileo: Mission to Jupiter

Galileo was one of the many scientific spacecraft delayed by *Challenger*'s 1986 loss. Launched in 1989, the *Galileo* spacecraft performed in-depth studies of Jupiter's atmosphere, moons, and its surrounding magnetic fields. One of Galileo's primary missions was the deployment of an atmospheric probe into Jupiter's atmosphere, accomplished successfully in 1995. After deploying the probe, *Galileo* itself went into orbit around Jupiter. Despite its high-gain antenna failure, about 70 percent of the mission's initial objectives were achieved—and the spacecraft is still returning images from Jupiter's moons in late 2001.

Cosmic Facts

The *Galileo* spacecraft was named after Galileo Galilei, the Italian Renaissance scientist who, using the newly invented telescope, discovered Jupiter's major moons in 1610.

On its way to Jupiter, *Galileo* flew by two asteroids, Gaspra in 1991 and Ida in 1993. These were our first close-up images of these fragments from the early solar system. *Galileo* was also the only vehicle in a position to take pictures of the far side of Jupiter when more than 20 fragments of Comet Shoemaker-Levy 9 plunged into Jupiter's atmosphere in July 1994.

Mars Pathfinder 1997

Finally breaking the Mars jinx, *Pathfinder* marked a NASA move to simpler, cheaper spacecraft for planetary exploration. Developed for a little over $200 million, *Pathfinder* used an innovative airbag landing system, and packed a mini-rover named Sojourner, which could extend the scientific reach of the lander.

Pathfinder landed on Mars and bounced to a safe stop on the Fourth of July 1997. Its landing site was a suspected outwash plain where water may once have flowed across the surface. Before it stopped transmitting data back to Earth in September, *Mars Pathfinder* returned 2.6 billion bits of information, including more than 16,000 images from the lander and 550 images from the rover, as well as more than 15 chemical analyses of rocks, and extensive data on winds and other weather factors.

The *Pathfinder* ended up operating for three times longer than expected. Scientific highlights of the *Mars Pathfinder* mission were …

- ◆ Martian dust includes magnetic particles.
- ◆ Rock chemistry at the landing site may be different from Martian meteorites found on Earth.
- ◆ Atmospheric clarity is higher than was expected.
- ◆ Evidence of wind abrasion of rocks and dune-shaped deposits was found, indicating the presence of sand.
- ◆ The measured atmospheric temperature profile was different than those derived from microwave measurements and Hubble observations.
- ◆ Rock size distribution was consistent with a flood-related deposit.
- ◆ Rounded pebbles and cobbles on the ground may mean that running water was present during a past epoch warm enough to keep liquid water stable.

Mars Global Surveyor '98

Mars Global Surveyor continued the run of good luck at Mars. The spacecraft left Earth on November 7, 1996, and dropped into orbit in September of 1997. MGS used the Martian atmosphere to drag it down into a near-circular mapping orbit around Mars, where it began its imaging mission in January 1998. Using a spare camera from the doomed *Mars Observer*, MGS has had a spectacular run of success in creating a global map of Mars that exceeds Viking orbiter resolution. The camera has produced more than 67,500 images and is still going strong. It is powerful enough to resolve details as small as four and a half feet.

Mars Global Surveyor quickly proved that the famous "face on Mars" was simply an eroded mesa, whose quirky shadows in Viking photos had produced the blurry impression of a human face. Mystery solved.

One of its most important discoveries is photographic evidence that shows liquid water may have burst from the Martian subsurface very recently. The fresh-looking mud and debris flows, seen on crater and valley walls, suggest that water may still be found very close to the frozen Martian surface. If water is still near the surface, the chances for life surviving there are much greater, and the water may also prove accessible to the first human explorers who visit Mars.

Mars Climate Orbiter and *Mars Polar Lander*

A new pair of probes was intended to join *Mars Global Surveyor* in 1999. The *Mars Climate Orbiter* was launched in December 1998. Its companion was the *Mars Polar Lander*. Both were designed to study Martian climate, its atmosphere, and surface conditions near the Martian south pole. Neither craft accomplished its mission, however.

Launched on a Delta rocket, the *Mars Climate Orbiter* was well on its way to becoming the first weather satellite around Mars. Just prior to Mars orbit insertion in September 1999, it streaked too low into the Martian atmosphere and burned up. NASA admitted that a "navigation error," stemming from a failure to convert metric to English units properly, put the craft too close to the atmosphere, and dragged it to its fate. Details count in spaceflight.

Mars Polar Lander was launched atop a Delta rocket on January 3, 1999, following the *Climate Orbiter* to Mars. After the embarrassing loss of the *Orbiter*, NASA had high hopes for its mission to sample the soil and weather conditions near the south pole. But the Mars jinx struck again—all contact was lost with the craft just before it entered the Martian atmosphere on December 3, 1999. The best guess as to what happened is that another design flaw was at fault: when the craft snapped open its landing legs just prior to touchdown, a faulty circuit sent a signal to the descent engines to shut down. The spacecraft probably dropped unpowered the last several hundred feet onto the surface and crashed.

Mars Odyssey

NASA cancelled the next lander mission to Mars while it reevaluated its design and testing practices. But the pull of Mars was too strong to stop our efforts to explore that world. At 11:02 A.M. on April 7, 2001, the *2001 Mars Odyssey* spacecraft left the Earth onboard a Delta II rocket from Cape Canaveral. About 53 minutes later, flight controllers at NASA's Jet Propulsion Laboratory received the first signal from the spacecraft through the Deep Space Network station in Canberra, Australia. *Mars Odyssey* was off to a perfect start.

The craft carries three scientific instruments designed to examine the Martian surface, to detect water and buried ice, and to characterize the radiation environment around it. An infrared imager will identify mineral types in the Martian soil and rocks, search for shallow ice deposits, and see if radiation levels are safe for human explorers. During its six-month cruise to Mars, the spacecraft turned on and calibrated its instruments. The spacecraft also fired its thrusters five times, fine-tuning its flight path to Mars.

> **Space Talk**
>
> **Aerobraking** is the use of the friction of the atmosphere to lower a spacecraft's orbit. Without aerobraking, the spacecraft would need to carry much more fuel. *Magellan* performed the first aerobraking experiments at another planet when it dipped into Venus's atmosphere in its final days in orbit in 1994.

The Odyssey arrived at Mars on October 24, 2001. After firing its main engine to slow itself enough to be captured by Mars's gravity, the orbiter at first circled the planet once every 25 hours. Over the following 76 days, the spacecraft edged closer to Mars, using *aerobraking*.

Once in Mars orbit, it began its mapping mission, which will last a full Martian year, or 29 Earth months. Later, it will serve as a communications relay for the next generation of Mars landers, due on the red planet in 2003.

Near-Earth Asteroid Rendezvous—*NEAR*-Perfect!

The smallest bodies in our solar system are the ones that happen to come closest to Earth—the near-Earth asteroids. These stadium-sized fragments of the early solar system sometimes come within the orbit of the Moon, and their mountain-sized cousins still pose a threat to our continued life on Earth.

The NEAR-Earth Asteroid Rendezvous mission—*NEAR*—was designed to visit the largest Earth-approaching asteroid, Eros. The spacecraft left Earth in February 1996 and cruised by asteroid Mathilde before heading for Eros. *NEAR* missed its first pass by Eros in late 1998 because of an engine malfunction, but was retargeted to arrive on Valentine's Day, 2000. The spacecraft dropped gingerly into an elliptical orbit around potato-shaped Eros, a 21-mile-long chunk of asteroidal debris about the size of Manhattan. Over the course of the next year it produced thousands of images of Eros' dusty surface, its boulders and craters.

Finally, on February 12, 2001, controllers at the Johns Hopkins Applied Physics Lab ended the *NEAR* mission with a gentle touch-down on Eros's surface. The orbiter was never designed to land, but it braked just enough with its last ounces of fuel to bump down onto the dusty surface at walking speed. The close-up images obtained during the descent, still under interpretation, will tell us much about how this ancient body has evolved in the 4 billion years it's been orbiting the Sun.

Genesis: Catching a Piece of the Sun

A solar-system-wandering spacecraft known as *Genesis*, which launched from Cape Canaveral during the summer of 2001, will capture pieces of the Sun during the autumn of that year and return them to Earth.

Our only solar samples come from the meteoritic dust, and meteorites themselves, that continually fall into Earth's atmosphere. All that material is missing some of the lightest, most volatile elements. *Genesis* is designed to capture and preserve the most pristine solar material flowing out through the solar system.

The mission is expected to capture about 10 to 20 micrograms of the solar wind, made up of charged atomic particles—like protons and helium nuclei—expelled by the Sun. *Genesis* will grab the samples during a two-year stay at a point about a million miles closer to the Sun than the planet Earth. The particles, about the weight of a few grains of salt, will be returned to Earth in a re-entry capsule in the summer of 2004, and the mission will end with a spectacular mid-air helicopter capture. Scientists will keep the pieces of the Sun in a special laboratory for study. The researchers hope to answer fundamental questions about the exact composition of our star and the birth of our solar system.

The Least You Need to Know

- ◆ Robotic spacecraft are the key to extending our senses into the distant universe.
- ◆ Space robots can send much more detailed info back from space than human eye-witnesses alone.
- ◆ The best robots still cannot answer complex questions about the presence of life on Mars; it will take a combination of people and machines to tackle those tough unknowns.
- ◆ NASA is conducting a vigorous investigation of many of the planets and asteroids of our solar system.
- ◆ We'll get our first unaltered samples of the Sun if the *Genesis* mission succeeds in capturing a bit of the solar wind.

30

Our Future in Space

In This Chapter

- ◆ Where from here?
- ◆ To look out or look in?
- ◆ Where did it begin?
- ◆ How will it end?

The last chapter usually means the end of the story, but this is just the jumping-off point for NASA and its forward-looking engineers, astronauts, and scientists as well as all those who share their dreams. I've talked in some detail about NASA's accomplishments during its more than 40-year existence. And I've looked at what NASA is up to today. But in this new century, one thing the space program looks forward to is change. In this chapter we'll take a look at some of the things that NASA may be involved in 10, 25, or maybe 50 years from now.

Beyond Low Earth Orbit

When I joined NASA as an astronaut candidate in 1990, my class adopted a shoulder patch that showed the space shuttle, the Moon, and Mars, and 23 stars, one for each member of the thirteenth group of astronauts. We fully expected that before 2000, we'd be headed back to the Moon, or even beyond near-Earth space to Mars or the asteroids. That reality hasn't materialized, but the reasons for going to those destinations are still valid.

For the last thirty years, our efforts in human space exploration have been restricted to low Earth orbit. We threw away the heavy-lift Saturn V booster in the early 1970s, and we've been flying the shuttle for the last 20 years. The Saturn V could put 110 tons into low Earth orbit (to kick its Apollo spacecraft off to the Moon), while the shuttle delivers about 18 tons of useful payload (and another 105 tons of orbiter, of course). So we've made a trade—reusability has replaced the capability to send humans out of low Earth orbit. The reason for the choice has been money.

NASA was able to go to the Moon because of Cold War competition with the Russians. Once we won the Moon race, America's (and the government's) attention turned elsewhere, and space was no longer the priority for federal funding. During Apollo we spent a peak of about 3.7 percent of the federal budget on space exploration. In 2001, we spent about 0.8 percent, or about $14 billion. The percentage has been steadily shrinking throughout the 1990s, as NASA's budget was cut every year since 1994. Only in 2001 has NASA's budget begun to tick upward, but it's barely keeping pace with inflation. Take a look at NASA's 1998 budget: $13.5 billion. By comparison, in 1998 Americans spent more than $30 billion on pizza (by the way, it's still impossible to get a decent pizza on the shuttle … the life of an astronaut is not an easy one!).

Let's look at NASA's 2001 budget compared to other federal expenditures: Social Security spending was 24.2 percent of the budget (30 times more than NASA's). Payments on the national debt took 20.7 percent of the federal budget—26 times the NASA total. Another 15.4 percent of the federal budget went to defense spending—almost 19 times the NASA percentage. And Health and Human Services spending was 22.7 percent of the budget—28 times greater than NASA's funding. I would argue that our wealthy country can afford a slightly bigger investment in the future. By increasing NASA's budget by 50 percent, we would provide enough money over ten years to once again leave low Earth orbit.

But if we had the gas money, where would we humans go beyond the space station? The time to start planning is now, because with the ISS nearing completion in four years, we will start to lose the corps of engineers and scientists that have been working hard on that project. It would be much wiser to keep that experienced talent pool together for journeys beyond low Earth orbit, rather than try to re-create it at great cost a few years down the road. So where should we plan to go?

Return to the Moon?

The most obvious destination is the Moon. The Apollo missions barely scratched the surface of exploring our companion satellite. Imagine trying to characterize the variety and complexity of Earth with a thousand pounds of rocks from just six sites around the globe.

Establishing an outpost on the Moon for rotating shifts of scientists would continue the process of unraveling the origin and history of that body; its surface still preserves events that have been obscured on Earth by our planet's dynamic surface processes. The Moon is also a superb site for deep-sky radio and optical astronomy—stable, interference-free, and blessed with no atmosphere. More promising still is the promise that water may be hidden at the lunar poles, a possible resource that would make supporting a lunar outpost much cheaper than one supplied only from Earth. The lunar soil is also known to harbor plentiful supplies of Helium-3, implanted over billions of years by the solar wind. Helium-3 is an isotope tailor-made for use in fusion reactors, producing very few radioactive byproducts in the fusion process. Fifty years down the line, we may be fueling our nuclear reactors with clean-burning nuclear fuel from the Moon.

In the near-term, the best reason for going back to the Moon, though, may be to use it as a testing ground for our journeys to Mars. Just three days away, the harsh lunar surface environment would teach us how to build and work with reliable machines, how to establish a self-sufficient outpost, how to deal with dust, temperature extremes, and lower gravity. By putting our robots and astronauts through the wringer on the Moon, we will gain valuable experience before we set off to a planet where a safe haven is at least six months away.

Visiting Asteroids

Between Mars and Jupiter lies the *asteroid belt*. Astronomers have catalogued the orbits of more than 8,000 asteroids, and many times that number remain undiscovered.

This ancient material, condensed out of the dust and gas of the early solar nebula, tried to coalesce into a planet, but the growth of giant Jupiter interrupted the process and prevented its formation. These leftovers, called asteroids, range in size from about 600 miles in diameter (a true protoplanet) down to house-sized boulders and smaller. Over the course of 4.5 billion years, Jupiter's gravity and collisions between the bodies have kicked thousands into orbits that come close to the Earth. These are called "Near-Earth Objects," or NEO's.

Although the combined mass of all the asteroids equals only about one percent of the weight of the

Space Talk

The **asteroid belt** is a doughnut-shaped region swirling with tens of thousands of irregularly shaped pieces of rock of varying sizes—pieces of debris left over from the formation of the planets.

Cosmic Facts

Of the combined mass of all the asteroids (about five percent of the Moon's mass), more than half makes up the largest asteroid, the one called Ceres.

Earth, the original mass of the material was no doubt larger. Much of it has been ejected from the solar system by Jupiter's gravity, or swept up by collisions between asteroids and the inner planets.

The NEO's are an inviting target for human exploration. Many of them approach Earth in orbits that are very similar to our home planet's. That means the amount of rocket fuel required to get to one of these cosmic splinters is small in comparison to other solar system destinations. For example, it takes about six kilometers per second (about 3.7 miles per second) of velocity change (from burning rocket fuel) to leave low Earth orbit (LEO) and land on the Moon. To get to some of the most accessible NEO's, the velocity change from LEO is much lower, in some cases as low as four to five km/sec. Better yet, because of the asteroid's weak gravity, leaving for home is very cheap in terms of rocket fuel.

Cosmic Facts

We already know of mission profiles that can take us to an NEO and back in six months, for less fuel than it takes to go to the lunar surface—one-way! Better asteroid targets are being found all the time.

There are three pay-offs from sending humans to asteroids. First, we'll get a science bonanza as scientist-astronauts select and return rocks and soil that date back to the very formation of the solar system. These materials are the building blocks of the planets! Second, we'll learn about the internal structure and properties of these bodies so that if we ever have to divert one from a collision course with Earth, we'll be able to act on a wealth of information. Third, and most important for our long-term presence in space, a good fraction of NEO's contain water and carbon, as well as rich iron-nickel ore, locked up in their rocks. The water alone is reason enough to begin mining asteroids: it's as valuable as pure gold in space, providing breathing oxygen and the most efficient combination of rocket propellants, hydrogen and oxygen.

Dr. Jones's Corner

I also see NEO's as perfect destinations for testing the waters of deep space. Before *Apollo 11* headed for the lunar surface, *Apollo 8* and *Apollo 10* had gone to lunar orbit on "shakedown cruises" of the Apollo spacecraft. Before we take off for Mars, we should give our new spacecraft a workout by visiting a near-Earth asteroid; some of the most approachable will let us get there and back in anywhere from one to six months. Sounds like a prudent course to follow, speaking as an astronaut with a personal interest in safety.

Should We Go to Mars?

Even if we focus at first on the Moon or the nearby asteroids, most of us find our gaze returning inevitably to Mars. It's a natural destination. The atmosphere there would protect an outpost from meteoroids and radiation. The water we see there (visible in the polar caps and in thin, high-altitude clouds), and hope to find beneath the surface, would provide a great resource for an outpost—life support oxygen and a ready supply of rocket fuel. And the reason for going there is to address the most important question we can answer by studying the planets—how did life arise in our solar system?

Is There Life?

We know from our *Mariner*, *Viking*, *Pathfinder*, and *Mars Global Surveyor* visits that ancient Mars was warmer and wetter than today's frigid desert of a planet. Life may have gotten started there, just as it found its footing on Earth. Chemical and microscopic evidence from a stony meteorite that traveled from Mars to Antarctica hints at bacterial life on Mars billions of years ago. If life did establish itself, perhaps the subsurface water we suspect is there still harbors primitive life forms today. There's no way to find out without teaming up humans and machines, and going after the answers ourselves. Robots will never be able to do the sophisticated searches, subsurface drilling, and complex laboratory work required to chase down the facts about life on Mars. The record of early life has been erased on Earth. Perhaps the knowledge we seek is waiting for us on the Red Planet.

We already have workable plans that could get us safely to and from the Martian surface. Engineer Bob Zubrin in 1990 proposed his "Mars Direct" plan, a mission profile that isn't fancy, but uses the resources of Mars to make the trip cheaper and more practical. His innovative ideas have sparked NASA's own planners to follow his lead, with a Mars mission now following this rough scenario:

- ◆ A robot cargo vehicle, mounting an Earth return spacecraft, would land on Mars and begin to manufacture rocket fuel for its empty tanks from the carbon dioxide in the Martian atmosphere.
- ◆ Two years later, with the return tanks full, a lander with the first human crew would touch down nearby, along with a second robot cargo ship. The explorers would erect an inflatable habitat and deploy rovers to expand their reach to surrounding areas.
- ◆ After 18 months on Mars, the crew would blast off in the return vehicle for Earth, leaving the outpost ready for the next crew arriving a few months later.

The total trip time would be about two to three years, and the outpost would be steadily expanded as a new human crew arrives every other year.

Spartan Approach

The details differ slightly as the plan evolves, but the common characteristics in this approach are …

1. The return vehicle gets fueled on Mars, avoiding the need to haul all that rocket propellant from Earth.

2. Every two years, a piloted ship and a robot return vehicle head for Mars, so that there is always a backup return vehicle on the surface in case the first doesn't function.

3. The outpost makes use of local resources—water, the atmosphere, and even rocks and soil—to provide life support, grow food, and construct shelter.

Cosmic Facts

Will Mars explorers be the first astronauts to carry on romantic relationships in space? Just as early explorers and colonists took their wives and families along, it'll be natural for humans to take their loves and loved ones along with them to Mars. (Have human beings had sex in space already? I can tell you with assurance that the answer thus far is "No!" although pregnancy tests are standard equipment for female astronauts who may have become pregnant on Earth but not realized it until in space.)

This spartan approach attacks the major objection to the human exploration of Mars—its cost. For an extra $3 billion per year (one-fifth of the current NASA budget), we could, over 10 years, pay for the first expedition, and subsequent ones every 2 years thereafter.

Unlike the Apollo expeditions, which ended abruptly after six voyages to the Moon, our presence on Mars should continue to expand with time. Each successive arrival would add to the self-sufficiency and safety of the new outpost, and if local resources can be used, our "Antarctic-style" camp on Mars could become a self-supporting colony. The main export? Knowledge. Mars is equal in surface area to that of all Earth's continents combined, so there are decades of exploration beckoning us on. We don't know where possible Martian fossils, or even hard-scrabble life forms, are hidden, and our early discoveries will only prompt more questions. Mars will furnish something very valuable to our human consciousness: an open frontier and a sense that we've yet to discover all that the universe holds in store for us.

Europa

While Venus and Mercury are too hostile for human visits, astronauts may one day attempt a journey to Jupiter's moon Europa, which likely harbors an ocean beneath its icy crust. The possibility that such an environment could support life is reason enough to pay a visit.

Trips to and beyond Jupiter will likely require development of nuclear thermal or nuclear-plasma propulsion systems; their higher operating temperatures permit much more efficient use of propellant, and shorter transit times to Mars and beyond. My astronaut colleague Franklin Chang-Diaz is working on just such an advanced propulsion system in his lab at the Sonny Carter Training Facility near JSC. Franklin's system, the variable specific impulse magnetoplasma rocket (VASIMR), uses a nuclear reactor to produce electricity and heat hydrogen. The electricity is used to heat the hydrogen to extremely high temperatures and ionize it into a plasma—a hot soup of charged particles, nuclei, and electrons. Using magnetic fields and radio waves, the engine confines the plasma and makes it very energetic (hot), and then releases the particles out a nozzle formed from magnetic field lines. The hot gas escapes at very high velocity—thus giving very high fuel efficiencies. The new rocket can also be used at lower efficiencies to generate high thrust. Trip times to Mars can be reduced from six to nine months using chemical fuels to about two to three months using this technology. Test flights are expected in two to three years.

Cosmic Facts

Next to Mars, Jupiter's moon Europa is another possible refuge for life in our solar system.

Other Solar Systems

Trip times may eventually limit the physical reach of human presence, but can't shackle our imaginations. And there is no reason why man's curiosity about other planets has to be limited to those of our solar system. As our ability to see great distances into space improves, astronomers are discovering more and more planets around other solar systems. Recently a team of astronomers has found a Jupiter-sized planet orbiting a faint nearby star similar to our Sun. The planet is the second object found orbiting the star 47 *Ursae Majoris* in the Big Dipper, also known as *Ursa Major* or the Big Bear.

Dr. Jones's Corner

Can astronauts withstand the long separations from Earth that a trip to Mars would require? You might think that, after spending six months in space in the International Space Station, an astronaut would be anxious to return to Earth. But that is not the case. Some astronauts, like Susan Helms, who lived for more than five months in the space station, find it hard to let go. "I think I've been in denial about this whole departure thing," Helms said only hours after leaving the space station. "I wasn't really quite ready to go because I've enjoyed the tour so much. As we were looking at [the space station] get smaller and smaller in the window [of the space shuttle], I think at least a couple of us felt a feeling of sadness that the whole adventure had come to an end." In the early '90's, one cosmonaut spent almost 14 months on Mir. A round trip to Mars, though longer, would certainly hold more interest and challenge than we've had to endure circling Earth. (At the moment, it takes anywhere from seven months to two years to get to Mars, depending upon how big a rocket is used and the specific type of spacecraft orbit followed to get there.) Humans will be able to survive such a trip—and do a fantastic job in the bargain.

The new planet is at least three-fourths the mass of Jupiter and orbits the star at a distance that, in our solar system, would place it beyond Mars but within the orbit of Jupiter. "Astronomers have detected evidence of more than 70 *extra-solar* planets," said Morris Aizenman, a senior science advisor at the National Science Foundation. "Each discovery brings us closer to determining whether other planetary systems have features like those of our own."

Space Talk

Extra-solar means belonging to a sun other than our own.

NASA plans a follow-on to the Hubble telescope, the next-generation space telescope, which will combine several large reflectors to form an optical interferometer—a telescope system that's much more powerful than the size of its individual light collectors. With this device located at a gravitationally stable point in the Earth-Moon system, we'll have the resolving power to see individual planets around nearby stars.

Measurements of the light they reflect may tell us if the chemicals in their atmospheres contain key indicators of life—water vapor and molecular oxygen. NASA's "TOPS" program (Toward Other Planetary Systems) is aimed at taking our imagination, and the boundaries of our knowledge, to other solar systems and their yet unknown worlds.

Searching for the Origin of the Universe

On July 1, 2001, a Delta rocket, blasting off from Cape Canveral, sent a satellite into orbit—a satellite designed to answer the most fundamental of questions. By searching for the oldest light in the universe, it would attempt to learn about the "Big Bang"—that is, the origin of the universe.

The satellite, which is called *MAP* (Microwave Anisotropy Probe), is being sent to a spot one million miles from Earth, on the side of our planet opposite from the Sun. There, its receivers will avoid the microwaves radiating from the Earth and Sun; this static is more than one billion times stronger than the signals the spacecraft will seek. The probe will construct four full-sky pictures of this so-called fossil light, each one taking six months. It will measure the slight temperature differences in the microwave background with an accuracy within one-millionth of a degree. The questions *MAP* hopefully will answer include:

♦ How old is the universe?

♦ What is the universe made of?

♦ How is the universe shaped?

♦ What is the universe's fate?

Types of Human Spaceflight

The twenty-first century astronaut is likely to look a little different than those who've flown in space over the last 40 years. NASA will still need a professional "Corps of Discovery," the spiritual descendants of Lewis and Clark, who will construct the space station and crew the next generation of vehicles bound for the Moon, the asteroids, and Mars. There will still be shuttle pilots and mission specialists, ISS Expedition Commanders and Flight Engineers. The ISS will see an increasing number of payload specialists operating specific experiments for a couple of weeks, then returning to Earth with the space shuttle or the Soyuz.

We'll no doubt see a growing number of space tourists—an exclusive club at first, but sure to become more commonplace as commercial spaceflight companies open their doors. Within ten years, we should expect to see industrial or commercial astronauts in space. These men and women will guide tourists on their journeys, operate commercial facilities joined to the ISS, or crew independent facilities that are lofted and supplied by commercial launch companies.

What about my next spaceflight? I'm still a little disappointed that my chance for a lunar voyage never materialized, but I'm nevertheless very satisfied with the privilege I've been given—to fly in space four times. I hope to make it back to orbit as a tourist in a few years. While I'm waiting to buy a ticket, I'll be content to look overhead every few nights and see the bright star of the ISS, with the U.S. Destiny lab anchoring the complex, soaring silently across the night sky. I feel the same pride a construction worker feels when he looks up at the skyscraper he helped top out. And when it comes time to go back to the Moon, to leave Earth orbit again—you can count me in.

> **Dr. Jones's Corner**
>
> I'm sure of one thing—whatever the reason for going into space, we'll have no shortage of volunteers.

The Least You Need to Know

- It's been nearly 30 years since any human has left Earth orbit.
- We possess the technology to go back to the Moon, or tackle voyages to the asteroids or Mars.
- The cost of going to Mars is affordable—spread over ten years, about the same cost as a major weapons system program.
- Our thirst for knowledge about the origins of life, and our place in the universe, means an expanding human presence in space is not a question of "if"—it's a question of "when."

Appendix A

Suggested Reading

Armstrong, Neil, Michael Collins, and Edwin E. Aldrin. *First on the Moon.* Boston: Little Brown, 1970.

Burrows, William. *This New Ocean: The Story of the First Space Age.* New York: Random House, 1998.

Carpenter, M. Scott, et al. *We Seven.* New York: Simon & Schuster, 1962.

Chapman, John L. *Atlas: The Story of a Missile.* New York: Harper & Brothers, 1960.

Clark, Phillip. *The Soviet Manned Space Program.* London: Salamander Books, 1988.

Clarke, Arthur C. *The Exploration of Space.* London: Temple Press, 1951.

Compton, W. David, and Charles D. Benson. *Living and Working in Space: A History of Skylab*, NASA SP-4208, Washington, D.C.: U.S. Government Printing Office, 1983.

Hacker, Barton C., and James M. Grimwood. *On the Shoulders of Titans: A History of Project Gemini.* NASA SP-4203, Washington, D.C.: U.S. Government Printing Office, 1977.

Hartt, Julian. *The Mighty Thor.* New York: Duell, Sloan and Pearce, 1961.

Heppenheimer, T.A. *Countdown: A History of Space Flight.* New York: John Wiley & Sons, 1997.

Lasby, Clarence G. *Project* Paperclip: *German Scientists and the Cold War.* New York: Atheneum, 1971.

Lay, Beirne. *Earthbound Astronauts: The Builders of Apollo-Saturn.* Englewood Cliffs, NJ: Prentice-Hall, 1971.

Ley, Willy. *Rockets, Missiles and Space Travel.* New York: Viking Press, 1957.

Lovell, Jim, and Jeffrey Kluger. *Lost Moon: The Perilous Voyage of* Apollo 13. New York: Houghton Mifflin, 1994.

Mullane, R. Mike. *Do Your Ears Pop in Space?* New York: John Wiley & Sons, 1997.

Murray, Charles, and Catherine Cox. *Apollo: The Race to the Moon.* New York: Simon & Schuster, 1989.

Oberg, James E. *The New Race for Space.* Harrisburg, PA: Stackpole, 1984.

Pierce, J.R. *The Beginnings of Satellite Communications.* San Francisco: San Francisco Press, 1968.

Rhea, John, ed. *Roads to Space: An Oral History of the Soviet Space Program.* New York: McGraw-Hill, 1995.

Richelson, Jeffrey. *America's Secret Eyes in Space.* New York: Harper & Row, 1990.

Smith, Melvyn. *Space Shuttle.* Newbury Park, Calif.: Haynes Publications, 1985.

Swenson, Loyd S., James M. Grimwood, and Charles C. Alexander. *This New Ocean: A History of Project Mercury.* NASA SP-4201, Washington, D.C.: U.S. Government Printing Office, 1966.

Walker, Martin. *The Cold War: A History.* New York: Henry Holt, 1993.

Winter, Frank. *Prelude to the Space Age: The Rocket Societies 1924-1940.* Washington, D.C.: Smithsonian Institution Press 1983.

Wolfe, Tom. *The Right Stuff.* New York: Farrar, Straus & Giroux, 1979.

Appendix B

Glossary

aerobraking The use of the friction of the atmosphere to lower a spacecraft's orbit. Without aerobraking, the spacecraft would need to carry much more fuel.

apogee The maximum altitude. In space flight, apogee means the maximum distance from earth that a spacecraft achieves during its orbit.

asteroid belt A doughnut-shaped region swirling with tens of thousands of irregularly shaped pieces of rock of varying sizes—pieces of debris left over from the formation of the planets.

astronaut The American term for one who travels in space.

atmosphere Layers of gases that surround a planet or one of its moons. Earth's atmosphere, commonly referred to as the air, is made up predominantly of nitrogen, oxygen, and carbon dioxide.

big bang theory Theory that states the universe was born with one gigantic explosion—thus explaining why everything is in motion and the universe is constantly growing.

black hole An infinitely dense object in space, with gravity so strong that even light cannot escape.

booster A rocket that gives a space vehicle its initial thrust to get off the launch pad. Usually discarded before reaching orbit.

canali Italian for channels, or grooves. It can also translate as canal, a man-made waterway built for transportation purposes.

capsule Spacecraft; usually a piloted one.

cohesive   Means that a substance has the tendency to stick to itself. You can make sand castles out of wet sand but not dry sand, for example, because wet sand is more cohesive.

cosmonaut   Russian term for space traveler.

countdown   The time sequence for launch preparations, counting backward with zero being the moment of a rocket's lift-off.

cryo   Short for cryogenic, or pertaining to substances at extremely low temperatures.

descent propulsion system   The main engine of the lunar module's descent stage, which lowered the lander to the Moon's surface. This descent engine could be throttled by the astronauts to enable the LM to hover.

dock   When two spaceships link themselves in space.

docking module   A special part of a space vehicle put in place so that it can dock and function as one with another spacecraft.

EVA (Extravehicular Activity)   A space walk. To venture outside of one's spacecraft—always, so far, for a constructive purpose.

extra-solar   Belonging to a sun other than our own.

flight plan   Detailed itinerary of events and tasks to execute during the course of a mission.

flyby   Term for the encounter that occurs when a space probe flies close enough to a planet to photograph it and gather other information, but not so close that it is captured into orbit by the planet's gravitational pull.

free fall   State of a body falling under gravity's influence alone. Weightlessness. Inside a spacecraft, because everything is falling at the same speed, it appears that nothing falls.

g-force   "One g" represents the normal force caused by gravity on our bodies while we are walking on Earth. When we are in an accelerating vehicle, we experience an increased force. That's why the back of your head presses harder against your seat when you accelerate in your car. Two g's would be twice the amount of pressure caused by gravity alone, four g's would be four times that force, and so on.

galaxy   A group of millions of stars bound together by gravity. Our Sun is a star in the galaxy known as the Milky Way.

gantry   The structure that stands next to a rocket, holds it in place, and allows men to work along the rocket's entire length by riding up and down in an elevator.

geostationary   An object that orbits at the exact speed of Earth's rotation, so that in that orbit it stays above the same location all the time.

geosynchronous relay   Transmitting a signal almost instantaneously around the world by sending it from the ground to a satellite in geosynchronous orbit around the Earth.

gravity   Force of attraction that objects with mass have toward one another, with the pull being stronger the heavier the object is. That is why the Earth has more gravity than the Moon, and the Moon has more gravity than an asteroid. Gravity is the force that keeps the planets in orbit around the Sun. It is the gravity of our Moon that creates tides on Earth.

heat shield   Protects the spacecraft from the flaming heat of re-entry.

hypervelocity   Loosely, anything going faster than Mach 5.

infrared   Light that has longer wavelengths than the visible radiation that the human eye can see.

intercontinental ballistic missile   A rocket capable of carrying a warhead from one continent to another.

internal pressurization   The force (in pounds per square inch) exerted by air that has been pumped into the space suit.

interstellar   Located or taking place among the stars.

IPT   Integrated Product Team. In the planning of a space mission, the IPT serves as a troubleshooting team that has the horsepower to make decisions and review the performance of all the key organizations. There are also smaller IPTs that support the Mission IPT: I was heavily involved with our EVA IPT, in charge of making sure all our space-walking tools and suit gear were ready in time for flight.

launch pad   Fireproof platform from which a rocket blasts off.

lift-off   The moment that a rocket leaves the Earth.

light year   The distance that light travels in a year. With light traveling at 186,000 miles per second, that's about six trillion miles.

lunar module   Called the LM or "Lem," the spacecraft used to land men on the Moon.

lunar rover   Specially made automobile used by astronauts to explore the lunar surface.

Manned Maneuvering Unit   A backpack device worn by early shuttle astronauts, equipped with small rocket thrusters. These rockets can be fired to move in the direction desired. The astronaut uses hand controls to choose the force and direction of the rockets, perhaps guiding himself to a nearby satellite to conduct repairs.

meteorology   The study and prediction of the weather.

Mir (pronounced *Meer*) Russian space station, launched in 1986, de-orbited in 2001.

mission specialists Astronauts, usually trained in science or engineering, who operate the space shuttle's science payloads, handle the robot arm, and don space suits to work outside the shuttle and station on space walks.

MOD Mission Operations Directorate.

moon A natural satellite of a planet, usually round. Mercury and Venus have no moons. Earth has one. Mars has two.

moon buggy *See* lunar rover.

NASA National Aeronautics and Space Administration, the overseeing body for American spaceflight since 1958.

nautical mile 6,076.1 feet, or 1.15 miles.

orbit The perpetual path of a planet around a sun, or the path of a moon around a planet. When a spacecraft circles a planet or moon it is said to be in orbit.

orbiter The reusable, winged portion of the space shuttle system, which actually orbits the Earth.

oxidizer Chemical that combines with the rocket fuel to produce the rapid burning in the engine's combustion chamber. The Saturn V used liquid oxygen because it was denser; more oxygen could be pumped into the tanks than if it was stored in gaseous form.

payload Cargo carried into space by a rocket.

perigee An orbit's closest approach to Earth.

planet Large celestial body in orbit around a sun.

radar altimeter Bounces radar signals off the terrain below to determine the height of the spacecraft above the surface. These measurements can be used to create a terrain map.

re-entry The mission phase when a spacecraft comes back into Earth's atmosphere after being in space.

resolution Refers to the smallest level of detail that can be distinguished, or "resolved," in an image. High-resolution photographs are more detailed and require either a low orbital altitude or a small telescope attached to the camera. A high-resolution camera zooms in on particular features, but can't cover the wide area of its low-resolution, or "low-rez," counterparts.

retro-rocket Rocket on a spacecraft which is fired to slow it down and make it re-enter the atmosphere.

satellite   An object which is in orbit around another object. The Earth is a satellite of the Sun and the Moon is a satellite of the Earth. *Sputnik I* was the first man-made satellite.

seas of the Moon   Dark plains sometimes referred to as "Maria"; these are not actually seas, and they never contained water. They are actually flat plains of ancient lava.

solar array wing   On Skylab, this was designed to gather up sunlight, and using photovoltaic cells, convert it into electrical power for use by the station.

solar panels   Flat panels that protrude from satellites. They gather rays of the sun and convert them into electricity.

sonic boom   Loud sound resembling the crackling of nearby thunder, caused when an object moving faster than the speed of sound causes a shock wave by piling up air in front of the vehicle.

space tourist   One who travels into space as a paying customer, just for the experience.

spectrograph   An instrument that splits light into very fine detail so we can identify specific atoms that are emitting or absorbing this light, thus giving us information about the specific energy state and composition of an object.

splashdown   The landing of a spacecraft in the ocean, using parachutes to ensure a soft landing.

Sputnik I   The first artificial satellite, built and launched by the U.S.S.R. Sputnik is Russian for "companion," as in "companion of the Earth."

supernova   A violent stellar explosion, the eventual fate of stars several times larger than our sun.

supersonic   Faster than the speed of sound.

thrust   Lifting force of a rocket.

thruster   Small rocket on a spacecraft used to adjust its position.

TIROS   Television InfraRed Observational Satellite, an early weather satellite.

trans-lunar injection   The firing of a spacecraft's engines to push it out of Earth orbit and onto a course for the Moon.

ufologist   An "expert" in the lore of unidentified flying objects.

ultraviolet   Light which has a wavelength so short that the human eye cannot see it.

warhead   A bomb attached to the nose of a rocket.

white dwarf   Final life stage of an average-sized star, in which it shrinks to a small, hot, white body.

white room   The small compartment on the gantry that fits around the spacecraft, providing a vestibule from which the astronauts enter the craft.

zero-gravity   Misnomer used to describe the sensation of no weight experienced in spaceflight. If you were precisely at the center of gravity of a spacecraft that was in perfect free-fall, you would perceive no external forces. But that doesn't happen very often—nearly always, astronauts live instead under conditions best described as microgravity, where forces experienced are very small compared to 1 g. These perceived forces are tiny compared to what we are used to, living on the surface of the Earth. This sensation of floating is called weightlessness or zero gravity.

Index

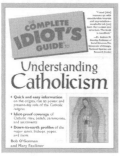

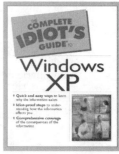